Elements of Solid State Physics

Elements of Solid State Physics

H. Y. Fan
Professor of Physics Emeritus
Purdue University

A WILEY-INTERSCIENCE PUBLICATION
JOHN WILEY & SONS
NEW YORK • CHICHESTER • BRISBANE • TORONTO • SINGAPORE

Library of Congress Cataloging in Publication Data:

Fan, H. Y.
Elements of solid state physics.

"A Wiley-Interscience publication."
Bibliography: p.
Includes index.
1. Solid state physics. I. Title.

QC176.F36 1988 530.4'1 87-12959
ISBN 0-471-85987-7

Printed in the United States of America

10 9 8 7 6 5 4 3 2 1

Preface

This book consists of organized lecture notes which were used many times for a one-year graduate course on solid state physics. It deals with crystalline solids, excluding the effects of contact with another medium. The properties of solids in contact with vacuum or another material are of great scientific and technological interest, including the case where a very thin layer of material is in contact with different materials on the two sides and may in a sense be considered two-dimensional. However, these properties belong to an area of specific problems of various complexities, for which the theories presented here are the basis of treatment.

The geometrically regular arrangement of atoms makes crystalline solids, three-dimensional and infinite in extent, the most easily comprehended model of condensed matter. Understanding gained from such idealized models is essential for the treatment of condensed matter in general, such as amorphous or glassy solids and liquid crystals.

This book is to be regarded as an introduction to the subject, being limited to considerations of basic problems. Its content is organized with an attempt to follow a logical sequence. Ion lattice symmetry and its direct consequences are treated in Chapter I. Following the adiabatic approximation, the treatment is divided into two parts: the system of ions, dealt with in Chapter II, and the system of electrons, treated in Chapters III and IV. Interactions of the two systems are considered in Chapter V. The extensive Chapter VI deals with the various properties of the solid, beginning with some general considerations in Section A and ending with Section F on structure and Section G on phase transitions.

There are a number of books on solid state physics and numerous review articles dealing with its various special topics. The bibliography lists some books which appear to be of broad interest.

As a lecturer, I have continually gained a clearer understanding of the subject from questions and discussions with students taking the course. In addition, I have benefited from discussions with many colleagues and research collaborators. It is a pleasure to acknowledge my indebtedness to all these friends.

H. Y. FAN

West Lafayette, Indiana
June 1987

Contents

Elements of Solid State Physics

CHAPTER ONE
Symmetry and the Lattice

1. SYMMETRY

The consideration of condensed matter is simplified if the substance possesses some symmetry in the spatial arrangement of ions, nuclei with firmly bound electrons. Symmetry refers to the fact that the matter appears the same from the point of view of a different coordinate system. An equivalent definition of symmetry is that the matter appears unchanged under an operation of a length-preserving, linear coordinate transformation in the fixed coordinate system. A linear transformation of Cartesian coordinates represented by a vector $\mathbf{x}$ to those represented by $\mathbf{x}'$ is given by

$$\mathbf{x}' = R\mathbf{x} + \mathbf{t}. \tag{1.1}$$

The transformation preserves length if the components of $\mathbf{t}$ are real and R is a real orthogonal matrix. A real orthogonal matrix has real components, and its inverse is its transpose. In a suitably oriented coordinate system, the matrix R has the form

$$R = \begin{vmatrix} \pm 1 & 0 & 0 \\ 0 & \cos\phi & -\sin\phi \\ 0 & \sin\phi & \cos\phi \end{vmatrix}. \tag{1.2}$$

R with $+1$ represents a rotation through the angle ϕ about the x_1 axis. R with -1 represents the same rotation followed by reflection across the x_2x_3 plane; the operation is called an improper rotation. R with $+1$ and $\phi = \pi$ is simply an inversion giving $\mathbf{x}' = -\mathbf{x}$. A symmetry operation may be represented by an operator

$$T \equiv \{R \mid \mathbf{t}\}, \tag{1.3}$$

which consists of an operation R followed by a translation $\mathbf{t}$.

Obviously, successive symmetry operations together are equivalent to some individual symmetry operation. A collection of distinct operations that are pure translations is a translation group or a group of primitive translations. A collection of operations $\{R \mid 0\}$ that do not involve any translation is a point group; all the rotation axes and reflection planes have a common point that remains fixed under all the transformations. A finite body can have only a point group of symmetry, and the point group usually refers to a body whose surface has the shape that gives the highest symmetry for a body of the given matter. The collection of all symmetry operations irrespective of the involvement of translation is the space group of the substance. Operations that include translation occur only for a body of infinite extent. The space group and point group of a substance apply to the bulk of a large body whose surface has negligible effects. In addition to pure translations, there are two more kinds of operations $\{R \mid \mathbf{t} \neq 0\}$: (1) rotation about an axis followed by a translation and (2) reflection across a plane followed by a translation. The axis is a screw axis, and the plane is a glide plane. These operations are sometimes included in the point group with their translations set to zero; it must be kept in mind, then, that the resulting "point group" contains modified transformations that are not symmetry operations.

Considerations of geometry show that there are 14 distinct translation groups, 32 point groups, and 230 space groups. There are 73 space groups that do not involve screw axes and glide planes; they are called symmorphic space groups. The number of symmorphic space groups (73) and the number of total space groups (230) are less than the product $14 \times 32 = 448$ because transformations about a point must hold about all points resulting from symmetry translations. Therefore each translation group is compatible with a limited number of point groups.

Certain substances may be considered two-dimensional or one-dimensional for the problems of interest. The number of relevant operations and consequently the number of distinct groups is therefore less than that for an actually three-dimensional substance. For the two-dimensional case, translations are limited to a plane, rotation axes can only be normal to the plane, and a reflection can only be across a line in the plane; there are 5 distinct translation groups, 10 point groups, and 17 space groups. One-dimensional substances have only two space groups.

2. ELEMENTS OF GROUP THEORY

In mathematics, a group is defined as a collection of elements $A, B, C, \ldots$ having the following properties:

(a) Any pair of elements combined in a prescribed way is also an element of the group.

(b) One of the elements, E, is a unit element or identity:

$$EP = PE = P, \tag{1.4}$$

where P is any element of the group.

(c) Every element P of the group must have an inverse P^{-1} that is an element of the group. In particular,

$$PP^{-1} = P^{-1}P = E. \tag{1.5}$$

(d) The associative law holds:

$$PQR = (PQ)R = P(QR). \tag{1.6}$$

Some examples of a group are:

1. The collection of all integers with addition as combination, $E = 0$, $P = n$, $P^{-1} = -n$.
2. The collection of $\exp(j/n)$s with multiplication as combination, where n is an integer and $j = 0, 1, \ldots, n-1$.
3. The collection of vectors $n_1\mathbf{a}_1 + n_2\mathbf{a}_2 + n_3\mathbf{a}_3$ with vector addition as combination, where n_1, n_2, n_3 are integers including zero.

Two special kinds of groups may be mentioned: Abelian and cyclic. A group is Abelian if the combination is commutative, that is, if $AB = BA$ for all the elements. In a cyclic group, all the elements are combinations of a single element. Obviously, a cyclic group must be Abelian. The three example groups given above are all Abelian. Only example 2 is a cyclic group.

The order of a group is the number of different elements in the group. A subgroup is a collection of some elements of a group that form a group by themselves; the identity element by itself is a trivial subgroup. For example, all even integers including zero constitute a subgroup of the group given above as example 1.

Coset. Let (H) be a subgroup, and let X be an element not contained in (H). $X(H)$ and $(H)X$ are, respectively, the left coset and right coset of group (G) under its subgroup (H). It can easily be shown that a left or right coset cannot have any element in common with (H). Furthermore, two right cosets or two left cosets either contain the same elements or have no common elements at all.

Expansion of a group in cosets. All the various right or left cosets together contain all elements of the group. Therefore a group can be expanded as

$$(G) = (H), X(H), Y(H), \ldots = [E, X, Y, \ldots](H) \tag{1.7}$$

or

$$(G) = (H), (H)P, (H)Q, \ldots = (H)[E, P, Q, \ldots]. \tag{1.8}$$

Index of a subgroup (H) in group (G). This is the ratio n/n_H, where n is the order of (G) and n_H is the number of elements in (H) and in each coset. The collection of elements $[E, X, Y, \ldots]$ or $[E, P, Q, \ldots]$ may not be a group. The two collections may not be the same.

Invariant subgroup. An invariant subgroup (H) satisfies the condition

$$(H) = X^{-1}(H)X, \tag{1.9}$$

where X is any element of the group. For such a subgroup,

$$X(H) = (H)X,$$

and expansions in right cosets and in left cosets are identical.

Factor group. An invariant subgroup and the cosets (right or left), each taken as an element, constitute the factor group:

$$(G/H) = [(H)], [X(H)], [Y(H)], \ldots. \tag{1.10}$$

It is easily shown that the factor group is indeed a group whose identity element is (H). The order of the factor group (G/H) is the index of (H) in (G).

Isomorphic groups. A group (G) is said to have an $n:1$ isomorphism with the group (g) if each element X_j of (G) corresponds to one element x_j of (g) but x_j corresponds to a collection of elements $X_{j1}, X_{j2}, \ldots, X_{jn}$ of (G). Two groups of $1:1$ isomorphism are said to be isomorphic or simply isomorphic groups. The elements of (G) that correspond to the identity element of (g) form an invariant subgroup of (G).

Direct-product group. Consider two groups

$$(G_1) = E, A_2, A_3, \ldots, A_i$$

and

$$(G_2) = E, B_2, B_3, \ldots, B_j$$

such that every element of one group commutes with all the elements of the other group. The direct-product group $(G_1 \times G_2)$ is defined by

$$(G_1 \times G_2) = E, A_2, \ldots, A_i, B_2, A_2B_2, \ldots, A_iB_2, \ldots, A_iB_j. \qquad (1.11)$$

It is easy to show that a product group is indeed a group.

Class of elements. Two elements, A and B, related by $B = X^{-1}AX$, where X is some element of the group, are said to conjugate to each other. Evidently, an element is conjugate to itself; if A is conjugate to B, then B is conjugate to A; if A is conjugate to B and C, then B and C are conjugate to each other.

A complete set of conjugate elements is a class of the group. The class containing an element A can be obtained by taking each element of the group to form an element conjugate to A. An element belongs to only one class; a group is divided into classes that have no elements in common. The identity element forms a class by itself, and therefore no other class is a subgroup. An invariant subgroup consists of complete classes in the group, including the class of the identity element.

Symmetry and group theory. The consideration of symmetry is covered by group theory. Symmetry operations are group elements, and successive operations are elements in combination. The translation group, point group, and space group in symmetry are groups in the sense of group theory. In terms of group theory, the translation group and point group are two subgroups of the space group, and a symmorphic space group is the direct-product group of the two.

The point group is isomorphic with the factor group. The translation group is an invariant subgroup, and it is Abelian.

Group multiplication table. For a group with noncommutative combination, such as some point groups of symmetry, it is expeditious to tabulate the combination for each pair of elements. The elements of the group are listed in the top row and in the left column. The element listed at the intersection of the mth row and nth column is the combination of the mth element in the column and the nth element in the row, with the latter preceding the first.

3. PROPERTIES OF MATTER AND SYMMETRY

A physical property P expresses the relation between two measurable quantities Q_1 and Q_2 of the substance: $Q_1 = PQ_2$. Q_1, Q_2, and P are given by tensors. The rank of P is equal to the sum of the ranks of Q_1 and Q_2. If P is the same for the quantities measured relative to two different sets of axes, then the transformation from one set of axes to the other is a symmetry element of the property P. The important and understandable Neumann's principle states that the symmetry of any physical property must include the symmetry elements of the point group of the substance. It should be pointed out that some properties possess certain inherent symmetries irrespective of the substance. For example, all second-rank tensor properties inherently have inversion symmetry. The symmetry possessed by a tensor reduces the number of independent components; the effect simplifies problems involving the tensor. In the transformation of x axes to x' axes with

$$x'_s = a_{sr} x_r, \tag{1.12}$$

the components of a tensor T transform according to

$$T_{i'j'\ldots} = (a_{i'l} a_{j'm\ldots}) T_{lm\ldots}. \tag{1.13}$$

The equations are written in the dummy suffix notation, and the range of each suffix is the three axes. A transformation of axes that is a symmetry element of property P gives

$$P_{ij\ldots} = P_{i'j'\ldots} = (a_{i'l} a_{j'm} \ldots) P_{lm\ldots}, \tag{1.14}$$

which provides a number of relations among the components of P, thereby reducing its number of independent components.

Sometimes symmetry requires some component of the property tensor to be symbolically different from itself. In such a case, the component must be numerically zero. For example, consider the component D_{133} of a third-rank property tensor, for example, a piezoelectric modulus. A rotation of axes by π about x_3 has

$$a_{1'1} = -1, \quad a_{2'2} = -1, \quad a_{3'3} = 1, \quad \text{and} \quad a_{r's} = 0 \quad \text{for} \quad r \neq s,$$

giving

$$D_{1'3'3'} = -D_{133}.$$

For a substance that contains such a rotation of axes in its point group of symmetry, we should have

$$D_{1'3'3'} = D_{133}.$$

Therefore, D_{133} must be zero for such substances. For the same reason, all the components that have only one 1 or one 2 in the subscript must be zero.

It should be borne in mind that the symmetry of a substance pertains to the substance in a definite condition that is specified by the so-called external parameters. Usually, the symmetry considered refers to a condition fully specified by two of the three external parameters temperature, pressure, and volume per unit mass. For many properties, however, the condition of the substance involves additional external parameters. For example, isothermal magnetoresistance and the Hall effect are electrical properties of a substance with an applied magnetic field as an additional external parameter, and electro-optical effects are optical properties of a substance with an applied electric field as an additional parameter. For such cases, the following evident principle is helpful: A substance with an additional external parameter possesses only those symmetry elements that are common to the substance without the additional external parameter and the external parameter by itself.

In the foregoing discussion about the symmetry of a property, property is defined as a relation between two measurable rather than measured quantities. The definition implies that the measurements of concern do not affect the symmetry. This is not always the case. For example, electrical resistivity relates the electric field and electric current density, and the measurements may be made at a field sufficiently high to significantly affect the symmetry of the substance. Should this be the case, the effect of the field on the symmetry of resistivity has to be taken into account like that of an additional external parameter.

4. LATTICES

A lattice consists of points in a regular geometrical arrangement. It is characterized by primitive translation vectors that number 3, 2, or 1 for a three-, two-, or one-dimensional lattice, respectively. The lattice is invariant under any translation that consists of multiples of individual primitive translation vectors. The primitive translation vectors define a primitive cell. Each primitive translation vector connects two neighboring points, each has a different direction, and no more than two vectors are coplanar. There is an arbitrariness in the choice of the vectors and the resulting shape of the primitive cell. However, a primitive cell always has the dimension ascribed to one point; that is, there are only points at the corners of the cell. A lattice of points is known as a Bravais lattice. The number of different Bravais lattices is the number of different groups of translational symmetry. Each Bravais lattice has the highest point-group symmetry compatible with the particular translational symmetry, which is called the holohedral point group.

A lattice is often described in terms of a unit cell, which has a more conventional shape. A unit cell may contain more than one primitive cell. It may be primitive (P), having points only at the corners; body-centered (I), with an extra point at the center; base-centered (C), with a point at the center of the base; or face-centered (F), with a point at the center of each surface. Lattices are classified into systems according to the shape of the unit cell. There are four systems for the five possible two-dimensional lattices:

oblique (P),
rectangular (P) and (C),
square (P), and
hexagonal (P).

The 14 possible three-dimensional lattices have seven systems:

triclinic	$a \neq b \neq c$, $\alpha \neq \beta \neq \gamma$, (P);
monoclinic	$a \neq b \neq c$, $\alpha = \beta = 90° \neq \gamma$, (P) and (C);
orthorhombic	$a \neq b \neq c$, $\alpha = \beta = \gamma = 90°$, (P), (C), (I), and (F);
tetragonal	$a = b \neq c$, $\alpha = \beta = \gamma = 90°$, (P) and (I);
cubic	$a = b = c$, $\alpha = \beta = \gamma = 90°$, (P), (I), and (F);
trigonal	$a = b = c$, $\alpha = \beta = \gamma < 120°$ ($\neq 90°$), (P); and
hexagonal	$a = b \neq c$, $\alpha = \beta = 90°$, $\gamma = 120°$ (P);

where a, b, c are axes of the unit cell $\alpha = (b,c)$, $\beta = (c,a)$, and $\gamma = (a,b)$.

A lattice of condensed matter is a network of basis ions. Basis ions are the smallest group of ions (nuclei with tightly bound electrons) with which the structure can be built up by translation. Depending on the structure of the unit of basis ions, the point-group symmetry of the lattice may be lower than that of the holohedral point group. On the other hand, the arrangement of ions in a primitive cell may introduce symmetry operations of screw axes and/or glide planes, operations consisting of a rotation or a reflection in combination with a translation shorter than a primitive translation, which are absent in a lattice of points. The space group describing the crystallographic structure of a substance gives the unit cell, the point-group symmetry, and the screw axes and glide planes that are present.

a. The Reciprocal Lattice

Let $\mathbf{a}_1$, $\mathbf{a}_2$, and $\mathbf{a}_3$ be the primitive translation vectors of a lattice. Those of the related reciprocal lattice are

$$\mathbf{b}_1 = \frac{\mathbf{a}_2 \times \mathbf{a}_3}{\mathbf{a}_1 \cdot (\mathbf{a}_2 \times \mathbf{a}_3)} \tag{1.15}$$

and similarly defined $\mathbf{b}_2$ and $\mathbf{b}_3$. It follows that

$$\mathbf{b}_i \cdot \mathbf{a}_j = \delta_{ij}. \tag{1.16}$$

The volume of a primitive cell is equal to $\mathbf{a}_1 \cdot (\mathbf{a}_2 \times \mathbf{a}_3)$ for the lattice and $\mathbf{b}_1 \cdot (\mathbf{b}_2 \times \mathbf{b}_3)$ for the reciprocal lattice.

The expression of position vector for a point of the reciprocal lattice is

$$\mathbf{B} = \beta_1 \mathbf{b}_1 + \beta_2 \mathbf{b}_2 + \beta_3 \mathbf{b}_3, \tag{1.17}$$

where β_1, β_2, and β_3 are integers. Let $\mathbf{B}_0$ be the position vector of the point closest to the origin along a particular direction. Then the three integers h, k, l in

$$\mathbf{B}_0 = h\mathbf{b}_1 + k\mathbf{b}_2 + l\mathbf{b}_3 \tag{1.18}$$

have no common denominator.

Consider now the direct lattice itself. The equation of a plane is

$$\mathbf{r} \cdot \mathbf{N}/N = d = \mathbf{R} \cdot \mathbf{N}/N,$$

where $\mathbf{N}$ is the normal from the origin to the plane and the constant d is the distance of the plane from the origin along the normal. $\mathbf{R}$ is the position vector of any one of the lattice points in the plane;

$$\mathbf{R} = \alpha_1\mathbf{a}_1 + \alpha_2\mathbf{a}_2 + \alpha_3\mathbf{a}_3,$$

where $\alpha_1, \alpha_2, \alpha_3$ are a set of integers. For an $\mathbf{N}$ equal to a $\mathbf{B}_0$ of the reciprocal lattice, we have

$$d = \frac{\mathbf{R}\cdot\mathbf{B}_0}{B_0} = \frac{\alpha_1 h + \alpha_2 k + \alpha_3 l}{B_0} \equiv \frac{n}{B_0},$$

where n is clearly an integer. According to the theory of numbers, there is always a set of integers $\alpha_1, \alpha_2, \alpha_3$ that gives $|n| = 1$ for a set of integers h, k, l having no common denominator. Therefore,

$$d_1 = 1/B_0 \tag{1.19}$$

is the spacing of parallel lattice planes (planes containing lattice points) perpendicular to $\mathbf{N} = \mathbf{B}_0$, one of which contains the lattice point at the origin. In view of the periodicity, d_1 is the separation between any neighbors among lattice planes perpendicular to $\mathbf{B}_0$.

The integers h, k, and l are called the Miller indices of lattice planes normal to $\mathbf{B}_0$. The Miller indices of a given lattice plane can be obtained from the intercepts $\gamma_1 a_2$, $\gamma_2 a_2$, and $\gamma_3 a_3$ of the plane on the axes of the three primitive translation vectors. Since

$$\cos(\mathbf{a}_1, \mathbf{B}_0) = \mathbf{a}_1\cdot\mathbf{B}_0/a_1 B_0 = h/a_1 B_0,$$
$$\cos(\mathbf{a}_2, \mathbf{B}_0) = k/a_2 B_0,$$
$$\cos(\mathbf{a}_3, \mathbf{B}_0) = l/a_3 B_0,$$

the normal from the origin to the plane has the length

$$d = \gamma_1 a_1 \cos(\mathbf{a}_1, \mathbf{B}_0) = \frac{\gamma_1 a_1 h}{a_1 B_0} = \frac{\gamma_2 a_2 k}{a_2 B_0} = \frac{\gamma_3 a_3 l}{a_3 B_0}.$$

Hence

$$h:k:l = \frac{1}{\gamma_1} : \frac{1}{\gamma_2} : \frac{1}{\gamma_3}. \tag{1.20}$$

The Miller indices h, k, l are the set of smallest integers that bear the ratios indicated by the $(1/\gamma)$s. The $(1/\gamma)$s are in general not integers.

For each direction of $\mathbf{B}_0$, there are points in the reciprocal lattice that have B's equal to various multiples of B_0. The point with $B = mB_0$ corresponds to geometrical planes in the direct lattice that have a separation of $1/mB_0$.

b. Brillouin Zones

A reciprocal lattice may be divided into zones (Brillouin zones) by drawing a bisector for each line that connects the origin with a lattice point. The zone containing the origin is the first zone. In order to reach a point in the ith zone from the origin, $i-1$ bisectors have to be crossed.

Two properties are important for the application of Brillouin zones:

(a) The volume of each zone is equal to the volume of the elementary cell of the reciprocal lattice. It should be noted that one zone may appear to consist of many adjoining parts; the first zone is an obvious exception.

(b) Any point can be brought into a zone or onto the zone boundary by applying a reciprocal-lattice vector **G**. There is one particular **G** in the first case. In the second case, there are at least two different **G**'s; that is, a point on the boundary of a zone differs from at least one other point on the boundary by some vector **G**.

5. LATTICE DETERMINATION

X rays have suitably small wavelengths for the investigation of the structure of matter. The Laue equations are derived from the consideration that the maximum intensity of the diffracted wave occurs under the condition that the waves from the scattering centers in the lattice have constructive interference; that is, they differ in phase by multiples of 2π. Let **S** be the vector bisecting the angle 2θ between the direction normals $\hat{N}_0$ and $\hat{N}$ of the incident and diffracted waves, respectively. The amplitude of **S** is

$$S = |\hat{N}_0 - \hat{N}| = 2\sin\theta. \tag{1.21}$$

Consider two scattering centers separated by

$$\mathbf{r} = \delta_1\mathbf{a}_1 + \delta_2\mathbf{a}_2 + \delta_3\mathbf{a}_3,$$

where $\delta_1, \delta_2, \delta_3$ are integers. The conditions for constructive interference of the waves scattered by the two centers are

$$\delta_1\mathbf{a}_1 \cdot \mathbf{S} = \delta_1 a_1 S\cos(\mathbf{a}_1, \mathbf{S}) = H\lambda,$$
$$\delta_2\mathbf{a}_2 \cdot \mathbf{S} = \delta_2 a_2 S\cos(\mathbf{a}_2, \mathbf{S}) = K\lambda,$$

and

$$\delta_3\mathbf{a}_3 \cdot \mathbf{S} = \delta_3 a_3 S \cos(\mathbf{a}_3, \mathbf{S}) = L\lambda,$$

where H, K, L are integers. For the constructive interference of scattering from various centers, similar conditions should hold for various δ_1's, δ_2's, and δ_3's, leading to the Laue equations:

$$\mathbf{a}_1 \cdot \mathbf{S} = a_1 S \cos(\mathbf{a}_1, \mathbf{S}) = h'\lambda = nh\lambda,$$
$$\mathbf{a}_2 \cdot \mathbf{S} = a_2 S \cos(\mathbf{a}_2, \mathbf{S}) = k'\lambda = nk\lambda,$$

and

$$\mathbf{a}_3 \cdot \mathbf{S} = a_3 S \cos(\mathbf{a}_3, \mathbf{S}) = l'\lambda = nl\lambda,$$

where h', k', l', n, h, k, l are integers and h, k, l have no integral divisor in common. These equations are satisfied with some λ. Then

$$\cos(\mathbf{a}_1, \mathbf{S}) : \cos(\mathbf{a}_2, \mathbf{S}) : \cos(\mathbf{a}_3, \mathbf{S}) = \frac{h}{a_1} : \frac{k}{a_2} : \frac{l}{a_3}. \tag{1.22}$$

By comparison with (1.20), we see that the ratios of cosines are the same as those of a lattice plane perpendicular to $\mathbf{B}_0 \parallel \mathbf{S}$ with Miller indices h, k, l. In this sense, a plane of Miller indices h, k, l may be considered a reflecting plane of the lattice for $\mathbf{S}$.

We substitute $\cos(\mathbf{a}_1, \mathbf{S})$ by $\cos(\mathbf{a}_1, \mathbf{B}_0)$ in (1.22), and note that $1/B_0 = d_1$ and $S = 2\sin\theta$. The first Laue equation becomes

$$2\sin\theta d_1 = n\lambda. \tag{1.23}$$

The same result is obtained from the other two Laue equations. This is Bragg's law. The integer n is called the order of reflection for the wavelength λ.

Ewald construction. It has been shown by Ewald that the condition for wave diffraction can be visualized clearly by means of a construction in the reciprocal lattice of the crystal. Referring to (1.21), we have

$$\hat{N}_0 \cdot \hat{s} = \hat{N}_0 \cdot (\mathbf{S}/\mathbf{S}) = -\sin\theta.$$

Bragg's law can be written as

$$\mathbf{k} \cdot \hat{s} \equiv \frac{\hat{N}_0}{\lambda} \cdot \hat{s} = -\frac{n}{2d_1}, \qquad \frac{\mathbf{k}}{2\pi}\left(\frac{\hat{s}n}{d_1}\right) = -\frac{(n/d_1)^2}{2}$$

where $\mathbf{k}$ is the wavevector of the incident wave. In the reciprocal lattice,

$$\mathbf{G} = n(\hat{s}/d_1)$$

is the position vector of the nth point along the direction $\hat{\mathbf{s}}$. Introducing $\mathbf{G}$, we get from Bragg's law

$$2(\mathbf{k}/2\pi)\cdot\mathbf{G} = -G^2, \qquad (\mathbf{k}/2\pi)\cdot(\mathbf{G}/G) = -G/2. \tag{1.24}$$

Let the vector $\mathbf{k}/2\pi$ be drawn in the reciprocal lattice, terminating at the origin. Draw a sphere containing the origin, with the center of the sphere at the initial point of $\mathbf{k}$. Any point of the reciprocal lattice that is found on the sphere has a $\mathbf{K}$ that satisfies Bragg's law. The direction of $\mathbf{G}$ gives the Miller indices of the reflecting planes for the wave of wavevector $\mathbf{k}$. The order number of the point along the direction of $\mathbf{K}$ from the origin is the order of reflection for the wavelength $\lambda = 2\pi/k$.

CHAPTER TWO
System of Ions

1. HAMILTONIAN OF A SUBSTANCE

Consider a region in a very large body, the region itself being sufficiently large to show the characteristic behavior of the substance. This region, sometimes referred to as the base region, it consists of many electrons and nuclei. Dynamically and for many types of behavior of the matter, each nucleus with some tightly bound electrons may be considered an entity, an ion. The Hamiltonian $\mathcal{H}$ of a substance is given then by the following expression:

$$\begin{aligned}\mathcal{H} = &-\sum_j \frac{\hbar^2}{2m}\nabla_j^2 - \sum_a \frac{\hbar^2}{2M_a}\nabla_a^2 + \sum_{j,k}{}' \frac{e^2}{2r_{jk}} \\ &+ V_{\mathrm{ei}}(\mathbf{r}_1,\mathbf{r}_2,\ldots,\mathbf{R}_1,\mathbf{R}_2,\ldots) + V_{\mathrm{ii}}(\mathbf{R}_1,\mathbf{R}_2,\ldots) \\ \equiv &\, H - \sum_a \frac{\hbar^2}{2M_a}\nabla_a^2.\end{aligned} \tag{2.1}$$

$\mathbf{r}$ and m are, respectively, the position vector and mass of an electron. Similarly, $\mathbf{R}$ and M are those quantities of an ion. Subscript a is a running index extending over all the ions. j and k in the subscript of $\mathbf{r}$ are each a running index covering all the electrons. V_{ei} is the potential of electron–ion interaction. $\mathbf{V}_{\mathrm{ii}}$ is the potential of ion–ion interaction.

The Schrödinger equation of the body is

$$\mathcal{H}\Phi = \mathcal{E}\Phi,$$

where $\mathcal{E}$ is the energy of the body. The so-called adiabatic approximation, first applied by Born and Oppenheimer, is to approximate the eigenfunction Φ by the following expression:

$$\Phi(\mathbf{r}_1,\mathbf{r}_2,\ldots,\mathbf{R}_1,\mathbf{R}_2,\ldots) = \psi(\mathbf{r}_1,\mathbf{r}_2,\ldots,\mathbf{R}_1,\mathbf{R}_2,\ldots)\chi(\mathbf{R}_1,\mathbf{R}_2,\ldots), \tag{2.2}$$

where ψ is the eigenfunction of H with $\mathbf{R}_1,\mathbf{R}_2,\ldots$ as parameters,

$$H\psi = E_R(\mathbf{R}_1,\mathbf{R}_2,\ldots)\psi. \tag{2.3}$$

Substituting the approximate Φ into the Schrödinger equation, we get

$$-\sum_a \frac{\hbar^2}{2M_a}\chi\nabla_a^2\psi - \sum_a \frac{\hbar^2}{M_a}(\nabla_a\psi)\cdot(\nabla_a\chi)$$
$$-\sum_a \frac{\hbar^2}{2M_a}\psi\nabla_a^2\chi + E_R(\mathbf{R}_1,\mathbf{R}_2,\ldots)\psi\chi = \mathcal{E}\psi\chi.$$

Multiplying this equation by ψ^* and integrating over $\mathbf{r}_1,\mathbf{r}_2,\ldots$, we obtain an equation for χ:

$$-\sum_a \frac{\hbar^2}{2M_a}\nabla_a^2\chi + E_R(\mathbf{R}_1,\mathbf{R}_2,\ldots)\chi$$
$$= \mathcal{E}\chi + \left[\sum_a \frac{\hbar^2}{2M_a}\int \psi^*\nabla_a^2\psi\, d\mathbf{r}_1\, d\mathbf{r}_2\ldots\right]\chi \tag{2.4}$$
$$+\sum_a \frac{\hbar^2}{M_a}\left(\int \psi^*\nabla_a\psi\, d\mathbf{r}_1\, d\mathbf{r}_2\ldots\right)\cdot(\nabla_a\chi).$$

The last two terms on the right-hand side can be shown to be negligible for the following two opposite, extreme cases: (a) The electrons behave as if they were perfectly free, and (b) the electrons are localized around individual ions. In case (a), the electron wavefunctions are independent of the ion coordinates. Hence $\nabla_a\psi$ and $\nabla_a^2\psi$ are zero, making the two terms vanish. Consider now case (b). ψ may be considered as a product of one-electron wavefunctions, each of the form $\varphi_2(\mathbf{r}_i - \mathbf{R}_a)$. Therefore,

$$\sum_a \nabla_a\psi = -\sum_i \nabla_i\psi.$$

For simplicity, let there be one type of ion. The coefficient of χ in the first term of interest becomes

$$\frac{m}{M}\sum_i \int \psi^* \left(-\frac{\hbar^2}{2m}\right) \nabla_i^2 \psi \, d\mathbf{r}_i,$$

which is $m/M < 1/1840$ times the kinetic energy of the electrons. The first term is therefore negligible in comparison with $\mathcal{E}\chi$. As to the second term, multiplication by χ^* and integration over $\mathbf{R}_a$ give

$$\sum_a \frac{\hbar^2}{M}\left(\int \psi^* \nabla_i \psi \, d\mathbf{r}_i\right) \cdot \left(\int \chi^* \nabla_a \chi \, d\mathbf{R}_a\right) = \sum_a \langle p_i \rangle \frac{\langle p_a \rangle}{M}.$$

In view of energy equipartition under equilibrium, we have

$$\overline{\langle p_i^2/m \rangle} \sim \overline{\langle p_a^2/M \rangle} \qquad \text{or} \qquad \langle p_i \rangle \sim (m/M)^{1/2} \langle p_a \rangle.$$

The result we obtained from the second term amounts to

$$\sim \left(\frac{m}{M}\right)^{1/2} \sum_a \frac{\langle p_a \rangle^2}{M},$$

which is about $(m/M)^{1/2}$ times the kinetic energy of the ions. Hence, the second term is also negligible.

Under the assumption that the two terms discussed are negligible, we have the generally used equation for ionic motion:

$$\left[-\sum_a \frac{\hbar^2}{2M_a} \nabla_a^2 + E_R(\mathbf{R}_1, \mathbf{R}_2, \ldots)\right] \chi = \mathcal{E}\chi. \tag{2.5}$$

E_R plays the role of potential energy. It includes the direct ion–ion interaction and the contribution of electrons with electron–ion interaction, for a set of $\mathbf{R}_1, \mathbf{R}_2, \ldots$; this is the essence of adiabatic approximation, which leads to a simplification in reasoning and in treatment.

2. LATTICE VIBRATION

In the discussion on lattice, it is stated that the lattice of a substance refers to identical groups of ions, basis ions, arranged in a regular pattern. The following notations will be used: $\mathbf{r}$ is the displacement of an ion from its equilibrium position, l is the index of a basis ion (one basis ion per primitive cell of the lattice), and b is the index of an ion in a basis ion. E_R in the equation of motion of the ions may be expanded in a series of ion displacements:

$$E_R = E_0 + \frac{1}{2}\sum_{lb,l'b'} \mathbf{r}_{lb} \cdot \mathbf{G}_{lb,l'b'} \cdot \mathbf{r}_{l'b'} + \cdots . \tag{2.6}$$

G is a second-rank tensor:

$$(G_{lb,l'b'})_{\alpha\beta} = \frac{\partial^2 E_R}{\partial(\mathbf{r}_{lb})_\alpha \partial(\mathbf{r}_{l'b'})_\beta} \qquad \alpha,\beta = 1,2,3. \tag{2.7}$$

There is no term linear in displacement $\mathbf{r}$, since E_0 for ions at their equilibrium positions should be the minimum of E_R. $E - E_0$ is the energy of oscillation of the ions. In the following development, terms of higher than the second order in displacement will be neglected. This approximation limits the treatment to harmonic motion of the ions.

The Hamiltonian of ion vibration is

$$H = \frac{1}{2}\sum_{l,b} \frac{1}{M_b}\mathbf{p}_{lb} \cdot \mathbf{p}_{lb} + \frac{1}{2}\sum_{lb,l'b'} \mathbf{r}_{lb} \cdot G_{lb,l'b'} \cdot \mathbf{r}_{l'b'}, \tag{2.8}$$

where $\mathbf{p}$ represents a momentum operator. There is the commutation relation

$$[\mathbf{r}_{lb}, \mathbf{p}_{l'b'}] = i\hbar I \delta_{ll'}\delta_{bb'}, \tag{2.9}$$

I being a unit tensor. Introduce two sets of new operators:

$$\mathbf{Q}_{\mathbf{q},b} = \frac{1}{N^{1/2}}\sum_{\mathbf{L}} \mathbf{r}_{lb} \exp(i\mathbf{q} \cdot \mathbf{L})$$

and

$$\mathbf{P}_{\mathbf{q}b} = \frac{1}{N^{1/2}}\sum_{\mathbf{L}} \mathbf{p}_{lb} \exp(-i\mathbf{q} \cdot \mathbf{L}). \tag{2.10}$$

N is the number of primitive cells in the region. The ion displacements may be regarded as a superposition of waves of displacement. Each wave is characterized by a wavelength λ and a wavevector $\hat{q}(2\pi/\lambda)$. $\mathbf{L}$ stands for $\mathbf{R}_l$. The reduced wavevector $\mathbf{q}$ is

$$\mathbf{q} = \hat{q}(2\pi/\lambda) - 2\pi\mathbf{G}, \tag{2.11}$$

where $\mathbf{G}$ is a reciprocal-lattice vector that leads to the smallest $\mathbf{q}$. We note that since the product of a lattice vector and a reciprocal-lattice vector is an integer, $\exp[i(\hat{q}/\lambda) \cdot \mathbf{L}] = \exp(i\mathbf{q} \cdot \mathbf{L})$, it is sufficient to consider the reduced wavevectors. The Fourier inversion theory gives the one-particle operators $\mathbf{q}$ and $\mathbf{p}$ in terms of the introduced operators $\mathbf{Q}$ and $\mathbf{P}$:

$$\mathbf{r}_{lb} = \frac{1}{N^{1/2}} \sum_{\mathbf{q}} \mathbf{Q}_{\mathbf{q}b} \exp(-i\mathbf{q} \cdot \mathbf{L})$$

and

$$\mathbf{p}_{lb} = \frac{1}{N^{1/2}} \sum_{\mathbf{q}} \mathbf{P}_{\mathbf{q}b} \exp(i\mathbf{q} \cdot \mathbf{L}). \tag{2.12}$$

The $\mathbf{Q}$'s and $\mathbf{P}$'s have a commutation relation similar to that of the r's and p's:

$$[\mathbf{Q}_{\mathbf{q}b}, \mathbf{P}_{\mathbf{q}'b'}] = i\hbar I \delta_{\mathbf{q}\mathbf{q}'} \delta_{bb'}. \tag{2.13}$$

$\mathbf{r}$ and $\mathbf{p}$ must be Hermitian, since they are operators of observables, and they must be real. From

$$\sum_{\mathbf{q}} \mathbf{Q}_{\mathbf{q}b} \exp(-i\mathbf{q} \cdot \mathbf{L}) = \sum_{\mathbf{q}} \mathbf{Q}^{+}_{\mathbf{q}b} \exp(i\mathbf{q} \cdot \mathbf{L})$$

and

$$\sum_{\mathbf{q}} \mathbf{Q}_{\mathbf{q}b} \exp(-i\mathbf{q} \cdot \mathbf{L}) = \sum_{\mathbf{q}} \mathbf{Q}^{*}_{\mathbf{q}b} \exp(i\mathbf{q} \cdot \mathbf{L}),$$

we get

$$\mathbf{Q}_{-q,b} = \mathbf{Q}^{*}_{\mathbf{q}b} = \mathbf{Q}^{+}_{\mathbf{q}b}.$$

Similarly,

$$\mathbf{P}_{-\mathbf{q},b} = \mathbf{P}^{*}_{\mathbf{q},b} = \mathbf{P}^{+}_{\mathbf{q}b}. \tag{2.14}$$

In terms of the $\mathbf{Q}$'s and $\mathbf{P}$'s, the Hamiltonian is

$$H = \frac{1}{2N} \sum_{\mathbf{L}b} \frac{1}{M_b} \sum_{\mathbf{q}\mathbf{q}'} \mathbf{P}_{\mathbf{q}'b} \cdot \mathbf{P}_{\mathbf{q}'b} \exp[i(\mathbf{q}+\mathbf{q}') \cdot \mathbf{L}]$$
$$+ \frac{1}{2N} \sum_{\mathbf{L}b\mathbf{L}'b'} \sum_{\mathbf{q}\mathbf{q}'} \mathbf{Q}_{\mathbf{q}b} \cdot G_{\mathbf{L}b\mathbf{L}'b'} \cdot \mathbf{Q}_{\mathbf{q}'b'} \exp[-i(\mathbf{q} \cdot \mathbf{L} + \mathbf{q}' \cdot \mathbf{L}')]. \tag{2.15}$$

Now,

$$\sum_{\mathbf{L}} \exp[i(\mathbf{q}+\mathbf{q}') \cdot \mathbf{L}] = \begin{cases} 0 & \text{for } (\mathbf{q}+\mathbf{q}') \neq 0 \\ N & \text{for } (\mathbf{q}+\mathbf{q}') = \mathbf{0}, \quad \text{i.e.,} \quad \mathbf{q}' = -\mathbf{q}. \end{cases} \tag{2.16}$$

The first term in the expression for H is therefore

$$\frac{1}{2}\sum_{\mathbf{q}b}\frac{1}{M_b}\mathbf{P}_{\mathbf{q}b}\cdot\mathbf{P}_{-\mathbf{q},b}. \tag{2.17}$$

Consider now the second term. It follows from the translation invariance of the lattice that

$$G_{lbl'b'} = G_{bb'}(\mathbf{L}-\mathbf{L}') \equiv G_{bb'}(\mathbf{D}). \tag{2.18}$$

Hence, the summation over $\mathbf{L}$ and $\mathbf{L}'$ can be replaced by a summation over $\mathbf{L}$ and $\mathbf{D}$:

$$\sum_{\mathbf{LD}} G_{bb'}(\mathbf{D})\exp(i\mathbf{q}\cdot\mathbf{D})\exp[-i(\mathbf{q}+\mathbf{q}')\cdot\mathbf{L}]$$
$$= N\sum_{\mathbf{D}} G_{bb'}(\mathbf{D})\exp(-i\mathbf{q}\cdot\mathbf{D}) \equiv NT_{bb'}(\mathbf{q}). \tag{2.19}$$

Therefore, the second term in the expression for H becomes

$$\tfrac{1}{2}\sum_{\mathbf{q}bb'}\mathbf{Q}_{\mathbf{q}b}\cdot T_{bb'}(\mathbf{q})\cdot Q_{-\mathbf{q},b'}.$$

The Hamiltonian of the substance may be written in the form

$$H = \sum_{\mathbf{q}} H_{\mathbf{q}}, \tag{2.20}$$

where

$$H_{\mathbf{q}} = \frac{1}{2}\left[\sum_{b}\frac{1}{M_b}\mathbf{P}_{\mathbf{q}b}\cdot\mathbf{P}^{+}_{\mathbf{q}b} + \sum_{bb'}\mathbf{Q}_{\mathbf{q}b}\cdot T_{bb'}(\mathbf{q})\cdot\mathbf{Q}^{+}_{\mathbf{q}b'}\right]$$

may be considered the Hamiltonian of the vibration wave with wavevector $\mathbf{q}$.

Consider $H_{\mathbf{q}}$ for a given $\mathbf{q}$. Bearing this in mind, we omit the index $\mathbf{q}$ in the subscripts of operators $\mathbf{Q}$ and $\mathbf{P}$. Express the operators in the following form:

$$\mathbf{Q}_b = \frac{1}{M_b^{1/2}}\sum_p B_p\mathbf{e}^{*}_{bp} \qquad \text{and} \qquad \mathbf{P}_b = M_b^{1/2}\sum_p A_p\mathbf{e}_{bp}, \tag{2.21}$$

with the condition

$$\sum_{b'}(M_bM_{b'})^{-1/2}T_{bb'}\cdot B^{+}_{p'}\mathbf{e}_{b'p'} = \omega^2_{p'}B^{+}_{p'}\mathbf{e}_{bp'}, \tag{2.22}$$

where ω_p is a number. The condition gives one vector equation for each ion b. A lattice with s different b's, s ions in a primitive cell, has s vector equations giving $3s$ solutions of ω_p^2, one for each degree of freedom of an ion b. Each solution is associated with a set of $B_p^+\mathbf{e}_{bp}$. Because the $T_{bb'}$'s are Hermitian ($T_{bb'} = T_{bb'}^+$), the equations have real roots ω_p^2, and the $\mathbf{e}_{bp}$'s obtained are orthogonal. We take normalized $\mathbf{e}_{bp}$ vectors:

$$\sum_b \mathbf{e}_{bp} \cdot \mathbf{e}_{bp'}^* = \delta_{pp'}. \tag{2.23}$$

Using the expressions of $\mathbf{Q}_b$ and $\mathbf{P}_b$, we get

$$\sum_{bb'} \mathbf{Q}_b \cdot T_{bb'} \cdot \mathbf{Q}_{b'}^+ \sum_p B_p \sum_b \mathbf{e}_{bp}^* \left(\sum_{p'} \omega_{b'}^2 B_{p'}^+ \mathbf{e}_{bp'} \right)$$
$$= \sum_p B_p \sum_{p'} \omega_{p'}^2 B_{p'}^+ \delta_{pp'} = \sum_p \omega_p^2 B_p B_p^+$$

and

$$\sum_b \frac{1}{M_b^{1/2}} \mathbf{P}_b \cdot \mathbf{P}_b^+ = \sum_{bpp'} A_p \mathbf{e}_{bp} \cdot A_{p'}^+ \mathbf{e}_{bp'}^*$$
$$= \sum_{pp'} A_p A_{p'}^+ \sum_b \mathbf{e}_{bp} \cdot \mathbf{e}_{bp'}^* = \sum_p A_p A_p^+, \tag{2.24}$$

where the A_p's and B_p's have the commutation relation

$$[B_p, A_{p'}] = i\hbar \delta_{pp'}. \tag{2.25}$$

We now have the following expression for $H_\mathbf{q}$:

$$H_\mathbf{q} = \sum_p \tfrac{1}{2} (A_{\mathbf{q}p} A_{\mathbf{q}p}^+ + \omega_{\mathbf{q}p}^2 B_{\mathbf{q}p} B_{\mathbf{q}p}^+). \tag{2.26}$$

It follows from the relations among the $\mathbf{Q}_{\mathbf{q}p}$'s and $\mathbf{P}_{\mathbf{q}p}$'s that

$$\mathbf{e}_{b,-\mathbf{q}p} = \mathbf{e}_{b,\mathbf{q}b}^*, \qquad B_{-\mathbf{q}p} = B_{\mathbf{q}p}^* = B_{\mathbf{q}p}^+,$$
$$A_{-\mathbf{q}p} = A_{\mathbf{q}p}^* = A_{\mathbf{q}p}^+. \tag{2.27}$$

$H_\mathbf{q}$ resembles the Hamiltonian of a group of independent harmonic oscillators. The transformation to operators $B_p \mathbf{e}_{bp}^*$ and $A_p \mathbf{e}_{bp}$ is analogous to the transformation to normal coordinates in the classical treatment of oscillations in a many-particle system. The ω_p's are frequencies of the normal modes.

The quantum-mechanical treatment of harmonic oscillators is facilitated by using the creation and annihilation operators, a^+ and a, of second quantization. Let

$$a_{\mathbf{q}p}^+ = (2\hbar\omega_{\mathbf{q}p})^{-1/2}(-iA_{\mathbf{q}p} + \omega_{\mathbf{q}p}B_{\mathbf{q}p}^+)$$

and

$$a_{\mathbf{q}p} = (2\hbar\omega_{\mathbf{q}p})^{-1/2}(iA_{\mathbf{q}p}^+ + \omega_{\mathbf{q}p}B_{\mathbf{q}p}). \tag{2.28}$$

These operators have the commutation relation

$$[a_{\mathbf{q}p}, a_{\mathbf{q}p}^+] = 1$$

and also the relation

$$a_{-\mathbf{q}p} = a_{\mathbf{q}p}^* = a_{\mathbf{q}p}^+. \tag{2.29}$$

In terms of these operators, we have

$$H_{\mathbf{q}} = \sum_p \hbar\omega_{\mathbf{q}p}(a_{\mathbf{q}p}^+ a_{\mathbf{q}p} + \tfrac{1}{2}). \tag{2.30}$$

The eigenvalue of $H = \sum_{\mathbf{q}} H_{\mathbf{q}}$ is given by

$$E = \sum_{\mathbf{p}}\sum_{\mathbf{q}} \hbar\omega_{\mathbf{q}p}(n_{\mathbf{q}p} + \tfrac{1}{2}). \tag{2.31}$$

$n_{\mathbf{q}p}$ is a positive integer or zero, and $\frac{1}{2}$ takes into account the zero-point energy. $\hbar\omega_{\mathbf{q}p}$ is a quantum of energy of the traveling wave $\mathbf{q}p$ of lattice vibration. It is called a phonon. $\mathbf{q}$ is the wavevector, p is referred to as a phonon branch, and there are $3s$ branches. An eigenstate of H is characterized by the numbers $n_{\mathbf{q}p}$ of the various kinds of phonons. The following useful results are well known:

$$a^+|n\rangle = (n+1)^{1/2}|n+1\rangle, \qquad a|n\rangle = n^{1/2}|n-1\rangle. \tag{2.32}$$

It is sometimes useful in the theoretical treatment of a problem to express operators in terms of $a_{\mathbf{q}p}^+$ and $a_{\mathbf{q}p}$. The following equations are easily deduced.

$$A_{\mathbf{q}p} = -i\left(\frac{\hbar\omega_{\mathbf{q}p}}{2}\right)^{1/2}(a_{-\mathbf{q},p} - a^{+}_{\mathbf{q}p}),$$

$$B_{\mathbf{q}p} = \left(\frac{\hbar}{2\omega_{\mathbf{q}p}}\right)^{1/2}(a_{\mathbf{q}p} + a^{+}_{-\mathbf{q},p}),$$

$$\mathbf{r}_{lb} = (NM_b)^{-1/2}\sum_{\mathbf{q}p}\left(\frac{\hbar}{2\omega_{\mathbf{q}p}}\right)^{1/2} \times \left[\mathbf{e}^{*}_{\mathbf{q}pb}\exp(-i\mathbf{q}\cdot\mathbf{L})a_{\mathbf{q}p} + \mathbf{e}_{\mathbf{q}pb}\exp(i\mathbf{q}\cdot\mathbf{L})a^{+}_{\mathbf{q}p}\right],$$

$$\mathbf{p}_{lb} = i\left(\frac{M_b}{N}\right)^{1/2}\sum_{\mathbf{q}p}\left(\frac{\hbar\omega_{\mathbf{q}p}}{2}\right)^{1/2} \times (\mathbf{e}_{\mathbf{q}pb}\exp(i\mathbf{q}\cdot\mathbf{L})a^{+}_{\mathbf{q}p} - \mathbf{e}^{*}_{\mathbf{q}pb}\exp(-i\mathbf{q}\cdot\mathbf{L})a_{\mathbf{q}p}). \tag{2.33}$$

Since the operators a and a^{+} have no diagonal matrix elements, $\mathbf{r}_{lb}$ and $\mathbf{p}_{lb}$ also have zero diagonal matrix elements or zero expectation values. On the other hand, only the diagonal matrix element of $r^2_{lb} = \mathbf{r}_{lb}\cdot\mathbf{r}_{lb}$ or p^2_{lb} is nonvanishing.

$$\langle|r^2_{lb}|\rangle = \left\langle\left|\frac{1}{NM_b}\sum_{\mathbf{q}p}\frac{\hbar}{2\omega_{\mathbf{q}p}}|\mathbf{e}_{b\mathbf{q}p}|^2(a_{\mathbf{q}p}a^{+}_{\mathbf{q}p} + a^{+}_{\mathbf{q}p}a_{\mathbf{q}p})\right|\right\rangle = \frac{1}{NM_b}\sum_{\mathbf{q}p}\frac{\hbar}{2\omega_{\mathbf{q}p}}2(n_{\mathbf{q}p} + \tfrac{1}{2})|\mathbf{e}_{b\mathbf{q}p}|^2. \tag{2.34}$$

3. PHONON SPECTRUM

In the simple case of one ion per primitive cell, there are one b and three phonon branches p. Equation (2.22) becomes

$$\frac{1}{M}\sum_{D} G_{ll'}(D)\exp(-i\mathbf{q}\cdot D)B^{+}_{\mathbf{q}p}\mathbf{e}_{\mathbf{q}p} = \omega^2_{\mathbf{q}p}B^{+}_{\mathbf{q}p}\mathbf{e}_{\mathbf{q}p}, \tag{2.35}$$

where the expression defining $T_{ll'}(\mathbf{q})$ has been introduced from (2.19). Now

$$\sum_{\mathbf{D}} G_{ll'}(\mathbf{D}) = \sum_{\mathbf{L}'} G_{ll'}(\mathbf{L} - \mathbf{L}') = 0$$

from the following consideration. Let the whole lattice be shifted by a small displacement $\mathbf{d}$. After the displacement has taken place, the ion at $\mathbf{L}$ experiences a force given by

$$\sum_{L'} G_{ll'}(\mathbf{L}-\mathbf{L}')\cdot\mathbf{d},$$

which should obviously be zero. Thus we conclude that

$$\omega_{\mathbf{q}p}=0 \qquad \text{for } \mathbf{q}=0. \tag{2.36}$$

The quantity $G_{ll'}(\mathbf{D})$, which represents the interaction of two ions separated by $\mathbf{D}$, decreases with increasing D. For small $\mathbf{q}$ such that $\mathbf{q}\cdot\mathbf{D}$ is small in the range of significant $\mathbf{D}$, we may expand $\exp(i\mathbf{q}\cdot\mathbf{D})$ contained in T_{bb} of (2.19). The term $-i\mathbf{q}\cdot\mathbf{D}$ makes no contribution to T_{bb} since it has opposite signs for $\mathbf{D}$ and $-\mathbf{D}$ whereas $G_{ll'}(\mathbf{D})$ does not change its sign. Using the next term in the expansion of $\exp(i\mathbf{q}\cdot\mathbf{D})$, we get from (2.35)

$$\omega_{\mathbf{q}p}\propto q/M^{1/2}. \tag{2.37}$$

The outlined frequency as a function of q characterizes the so-called acoustical branches. Acoustic waves belong to these branches, and the wavevectors are small in the sense discussed.

The direction $\mathbf{e}_{qp}$ of vibrational displacement is determined from the actual solution of (2.35) with the particular $T_{bb}(\mathbf{q})$. Sometimes symmetry considerations may be helpful in the determination of $\mathbf{e}_{\mathbf{q}p}$'s. For $\mathbf{q}$ along an axis of a cubic lattice, one $\mathbf{e}_{\mathbf{q}p}$ is evidently parallel to $\mathbf{q}$ and the other two are perpendicular to $\mathbf{q}$. Such vibrations are referred to as longitudinal and transverse modes of vibration, respectively. The terminology is widely used to indicate whether the vibration is more longitudinal or more transverse with respect to the propagation direction of the wave. In the special case under consideration, the two branches of transverse modes must be degenerate in frequency, and their directions of vibration $\mathbf{e}$ are not unique.

Consider now a lattice with more than one ion per primitive cell. There are $3s$ branches of vibration, s being the number of ions per primitive cell. Three of the branches are acoustical; the ions in a primitive cell tend to vibrate in unison for small q's or long wavelengths, and $\omega\to 0$ for $q\to 0$ as in a lattice that has one ion per primitive cell. For the other $3s-3$ branches, the ions in one primitive cell vibrate differently for small q, and $\omega\neq 0$ for $q=0$. These branches are often called optical branches. Electromagnetic radiation in the optical range has significant ω and small q, in the sense discussed here.

a. Density of Modes and Frequency Distribution

In order for it to be possible to represent a substance by a limited base region, the region should be sufficiently large that the phenomena of interest may be considered to be periodic from region to region. This is in essence the Born–von Kármán boundary condition. Let N_1, N_2, N_3 be the number of cells in the base region along the three primitive translation vectors $\mathbf{a}_1, \mathbf{a}_2, \mathbf{a}_3$. According to the boundary condition, a wave should have the same phase at opposite boundaries of the region. Therefore the wavevector $\mathbf{q}$ has to satisfy the condition

$$\mathbf{q} \cdot N_i \mathbf{a}_i = 2\pi n_i, \tag{2.38}$$

where the subscript i stands for 1, 2, or 3 and the n_i's are integers. Let $\mathbf{b}_1, \mathbf{b}_2, \mathbf{b}_3$ be the vectors defined by (1.15), and express the wavevector in terms of these vectors:

$$\mathbf{q} = \beta_1 \mathbf{b}_1 + \beta_2 \mathbf{b}_2 + \beta_3 \mathbf{b}_3.$$

It follows that

$$\beta_i = 2\pi n_i / N_i. \tag{2.39}$$

Since the n's must be integers, the $\mathbf{q}$'s are discrete, and each $\mathbf{q}$ occupies a space of

$$\frac{(2\pi)^3}{N_1 N_2 N_3} \mathbf{b}_1 \cdot (\mathbf{b}_2 \times \mathbf{b}_3) = \frac{(2\pi)^3}{N} \frac{1}{v} = \frac{(2\pi)^3}{V}, \tag{2.40}$$

where v is the volume of a primitive cell, N is the number of cells in the base region, and V is the volume of the base region. The number of discrete $\mathbf{q}$'s per unit volume of $\mathbf{q}$ space, $V/(2\pi)^3$, is the density of modes in the $\mathbf{q}$ space.

Let $\Omega_p(\omega)$ be the volume enclosed by the surface of constant frequency ω in the $\mathbf{q}$ space for the branch p. The number of discrete $\mathbf{q}$'s per unit frequency is

$$\rho_p(\omega) = \frac{V}{(2\pi)^3} \frac{d\Omega_p(\omega)}{d\omega}. \tag{2.41}$$

The reduced wavevector $\mathbf{q}$ is limited to a space of $(2\pi)^3 \times$ (volume of a Brillouin zone)$= (2\pi)^3/v$. The total number of discrete $\mathbf{q}$'s in a branch is

$$(V/8\pi^3)(8\pi^3/v) = N. \tag{2.42}$$

The number of **q**'s per unit range of ω, as a fraction of the total number of **q**'s in the branch, is

$$f_p(\omega) = \frac{\rho_p(\omega)}{N} = \frac{v}{(2\pi)^3}\frac{d\Omega_p(\omega)}{d\omega}. \tag{2.43}$$

$f_p(\omega)$ is called the frequency distribution function of phonons of the branch.

The derivative $d\Omega_p(\omega)/d\omega$ in the expressions for $\rho_p(\omega)$ and $f_p(\omega)$ requires knowledge of $\omega(\mathbf{q})$ that can be obtained only by solving Eq. (2.22) for the particular substance. The following general observations can be made.

1. Acoustical branches near $q = 0$. We have

$$\omega_{\mathbf{q}p} = c_p\mathbf{q} \tag{2.44}$$

for each branch, c_p being the proportionality constant of the branch. A surface of constant ω is spherical in **q** space. We have

$$\Omega_p(\omega) = \frac{4\pi}{3}\left(\frac{\omega}{c_p}\right)^3 \quad \text{and} \quad f_p(\omega) = \frac{v}{2\pi^2}\left(\frac{\omega^2}{c_p^3}\right). \tag{2.45}$$

2. Singularities in frequency distribution. An extremum of $\omega(\mathbf{q})$ gives rise to a singularity in the frequency distribution. In addition, saddle points of $\omega(\mathbf{q})$ occur. A saddle point in **q** space is the location where a surface of constant frequency constricts to a point. Saddle points are classified into two kinds, depending upon whether surfaces of higher or lower frequencies are broken into two parts. It has been shown by Van Hove that any periodic function, $\omega(\mathbf{q})$ being such a function, has at least a certain number of saddle points, which depends on the number of independent variables. In a three-dimensional lattice, there is at least one saddle point of each kind. Each saddle point produces a sharp kink in $f(\omega)$, that is, a discontinuity in $df(\omega)/d\omega$. Kinks of opposite sense are produced by two different kinds of saddle points. The above considerations show that the frequency distribution $f(\omega)$ is not actually a smooth curve.

4. LATTICE SPECIFIC HEAT

We take the specific heat c_v at constant volume to be the heat capacity of a unit mass of the substance at constant volume:

$$c_v = (\partial E/\partial T)_v, \tag{2.46}$$

where E is the internal energy. Classically, according to the equipartition theorem, each degree of freedom contributes to E an amount kT. Therefore, $E = 3NkT$, where N is the number of ions per unit mass. This leads to

$$c_v = 3Nk,$$

the classical law of Dulong and Petit.

According to the quantum-mechanical treatment of lattice vibration, the energy eigenvalues of the $\mathbf{q}p$ mode are given by

$$(\epsilon_{\mathbf{q}p})_n = (n + \tfrac{1}{2})\hbar\omega_{\mathbf{q}p}. \tag{2.47}$$

The probability for a mode to be in the state n is proportional to $\exp(-\epsilon_n/kT)$. Therefore the average energy of a mode at temperature T is

$$\begin{aligned}\langle\epsilon_\omega\rangle &= \frac{\sum_n \epsilon_n \exp(-\epsilon_n/kT)}{\sum_n \exp(-\epsilon_n/kT)} = \frac{\sum_n (n\hbar\omega)\exp(-n\hbar\omega/kT)}{\sum_n \exp(-n\hbar\omega/kT)} + \frac{1}{2}\hbar\omega \\ &= \left(\frac{1}{\exp(\hbar\omega/kT) - 1} + \frac{1}{2}\right)\hbar\omega.\end{aligned} \tag{2.48}$$

Hence the total energy of vibration is

$$E = \sum_p \int_{(\omega_1)_p}^{(\omega_2)_p} \langle\epsilon_\omega\rangle \rho_p(\omega)\, d\omega,$$

which leads to the following expression for the specific heat:

$$c_v = \sum_p \int_{(\omega_1)_p}^{(\omega_2)_p} k\left(\frac{\hbar\omega}{kT}\right)^2 \frac{\exp(\hbar\omega/kT)}{[\exp(\hbar\omega/kT) - 1]^2}\rho_p(\omega)\, d\omega. \tag{2.49}$$

The integration for each branch p covers the range from its lowest frequency $(\omega_1)_p$ to the highest frequency $(\omega_2)_p$. Usually, kT is very small in comparison with $\hbar\omega_1$ for any optical branch. The modes of these branches have very little temperature dependence in $\langle\epsilon_\omega\rangle$ and make negligible contributions to c_v. The Dulong–Petit law for c_v can be justified only at very high temperatures. With $kT > \hbar\omega_2$ for all branches,

$$c_v \to k\sum_p \int_{(\omega_1)_p}^{(\omega_2)_p} \rho_p(\omega)\, d\omega = 3Nk.$$

Such a high temperature is not realistic in general.

Two theories of specific heat of lattice vibration are well known. Einstein's theory assumes that all modes have the same frequency. A more refined assumption is made in Debye's theory:

$$\omega(\mathbf{q}) = cq. \tag{2.50}$$

Let the velocity constant be c_l for longitudinal waves and c_t for the two kinds of transverse waves. Then

$$\sum_p \rho_p(\omega) = \frac{1}{2\pi^2}\left(\frac{1}{c_l^3} + \frac{2}{c_t^3}\right)\omega^2 \equiv \frac{3}{2\pi^2 c^3}\omega^2 \tag{2.51}$$

per unit volume. Define a maximum frequency ω_m, which is assumed to be the same for the three kinds of waves,

$$\omega_m = c(6\pi^2 N)^{1/3}, \tag{2.52}$$

and a Debye temperature Θ,

$$\Theta/T = \hbar\omega_m/kT. \tag{2.53}$$

The expression for the specific heat given by the outlined approximations is

$$c_v = 9Nk(T/\Theta)^3 \int_0^{\Theta/T} \frac{e^x x^4\, dx}{(e^x - 1)^2}. \tag{2.54}$$

According to this expression, c_v increases with temperature. For $(\Theta/T) \ll 1$, the integral becomes approximately $(\Theta/T)^3/3$, and c_v approaches the saturation value $3Nk$, the Dulong–Petit value expected to be correct at high temperatures. For $(\Theta/T) \gg 1$, the integral is approximately $4\pi^4/15$, and

$$c_v = (12\pi^4 Nk/5)(T/\Theta)^3. \tag{2.55}$$

Usually, experimental data on $c_v(T)$ are considered in terms of the Debye theory. At sufficiently low temperatures, the observed specific heat shows the T^3 temperature dependence and yields a Θ that corresponds to a reasonable value of c. It is difficult to estimate precisely the effect of the approximation that c is independent of the direction of $\mathbf{q}$. At higher temperatures where larger ω's are significant, extrema and singularities of $\omega(\mathbf{q})$ may cause $c_v(T)$ to differ from a smooth curve. A smooth curve of $c_v(T)$ is often fitted with a temperature-dependent Debye temperature.

5. SCATTERING OF WAVES BY LATTICE VIBRATION

Consider a plane wave of electromagnetic radiation, $A\exp(i\mathbf{k}_0 \cdot \mathbf{r})$, that impinges on an atom situated at the origin. Under the assumption that the position and the state of the atom are fixed, the radiation is scattered coherently; that is, the scattered wave is in phase with the primary wave. A component of the scattered wave has the amplitude

$$A_s = (\hat{e}_s \cdot \hat{e}) F \frac{A}{R} e^{ikR} \tag{2.56}$$

at a sufficiently large distance R from the origin. F is the form factor of the scattering atom. $\hat{e}$ and $\hat{e}_s$ are the polarization vectors of the incident and scattered waves, respectively. $k = k_0$, but $\mathbf{k} \neq \mathbf{k}_0$. For an identical atom located at $\mathbf{R}_c$ from the origin, a factor $\exp(i\mathbf{k}_0 \cdot \mathbf{R}_c)$ should be added for the incident wave and a factor $\exp(i\mathbf{k} \cdot \mathbf{R}_c)$ should be subtracted for the scattered wave. Hence, the factor $\exp(i\mathbf{K} \cdot \mathbf{R}_c)$ has to be added to the above expression for the atom at $\mathbf{R}_c$, where

$$\mathbf{K} \equiv \mathbf{k}_0 - \mathbf{k}. \tag{2.57}$$

The amplitude of a wave scattered by a lattice of stationary ions is

$$A_s = (\hat{e}_s \cdot \hat{e}) \frac{A}{R} e^{ik'R} \sum_b F_b \sum_l \exp(i\mathbf{K} \cdot \mathbf{R}_{lb}), \tag{2.58}$$

where l is the index of the primitive cell, b is the index of an ion in a cell, and $\mathbf{R}_b$ is the position vector of the ion (l, b). We see that A_s vanishes unless $\mathbf{K}$ is a reciprocal lattice vector $\mathbf{G}$. Nonvanishing superposition or constructive interference is the basis of the previously discussed Laue equations and Bragg's law for x-ray diffraction. It should be noted that R in the denominator of expression (2.58) should actually vary somewhat from ion to ion. This complication has been neglected on the grounds that R is large in comparison with the maximum distance between ions that need to be considered.

Consider now the effect of ion vibration, the displacement of each ion from its equilibrium position $\mathbf{R}$ being small in comparison with the interionic distances. We then have

$$A_s \propto A e^{ik'R} \sum_l \sum_b F_b \exp(i\mathbf{K} \cdot \mathbf{R}_{lb}) \exp(i\mathbf{K} \cdot \mathbf{r}_{lb}). \tag{2.59}$$

The displacement of $\mathbf{r}$ of an ion being the sum of displacements $\mathbf{r}_{\mathbf{q}p}$ associated with various normal modes $\mathbf{q}p$ of vibration, we have

$$\exp(i\mathbf{K}\cdot\mathbf{r}_{lb}) = \prod_{\mathbf{q}p}\exp[i\mathbf{K}\cdot(\mathbf{r}_{lb})_{\mathbf{q}p}]$$
$$= \prod_{\mathbf{q}p}\{1 + i\mathbf{K}\cdot(\mathbf{r}_{lb})_{\mathbf{q}p} - \tfrac{1}{2}[\mathbf{K}\cdot(\mathbf{r}_{lb})_{\mathbf{q}p}]^2 + \cdots\}.$$

The ion displacements of interest are small compared with interionic distances. For the waves usually considered, $\mathbf{K}\cdot\mathbf{r} \ll 1$. Therefore, terms beyond the second order may be neglected. Then we get the following approximation:

$$\exp(i\mathbf{K}\cdot\mathbf{r}_{lb}) \approx \prod_{\mathbf{q}p}\{1 - \tfrac{1}{2}[\mathbf{K}\cdot(\mathbf{r}_{lb})_{\mathbf{q}p}]^2\} + \sum_{\mathbf{q}p} i\mathbf{K}\cdot(\mathbf{r}_{lb})_{\mathbf{q}p} \times \prod_{\mathbf{q}'p'\neq\mathbf{q}p}\{1 - \tfrac{1}{2}[\mathbf{K}\cdot(\mathbf{r}_{lb})_{\mathbf{q}'p'}]^2\}, \tag{2.60}$$

since $\mathbf{K}\cdot(\mathbf{r}_{lb})_{\mathbf{q}p} \ll \{1 - \frac{1}{2}[\mathbf{K}\cdot(\mathbf{r}_{lb})_{\mathbf{q}p}]^2\}$. The first term is responsible for elastic scattering, and the second term gives rise to inelastic scattering.

a. Elastic Scattering

Mathematically,

$$\lim_{N\to\infty}\prod_{n=1}^{N}\left(1 - \frac{a_n}{N}\right) = \exp\left(-\lim_{N\to\infty}\frac{1}{N}\sum_{n=1}^{N} a_n\right).$$

Therefore,

$$\prod_{\mathbf{q}p}\left\{1 - \tfrac{1}{2}[\mathbf{K}\cdot(\mathbf{r}_{lb})_{\mathbf{q}p}]^2\right\} \sim \exp\left\{-\tfrac{1}{2}\left[\sum_{\mathbf{q}p}\mathbf{K}\cdot(\mathbf{r}_{lb})_{\mathbf{q}p}\right]^2\right\}. \tag{2.61}$$

According to (2.34), the expectation value of $(\mathbf{r}_{lb})^2_{\mathbf{q}p}$ is independent of l. Furthermore, the angle between $\mathbf{K}$ and $(\mathbf{r}_{lb})_{\mathbf{q}p}$ is independent of l. The scattering given by the first term of (2.60) is

$$A_{s,e} \propto (\hat{e}_s\cdot\hat{e})Ae^{ik'R}\sum_b e^{-W_b}F_b\sum_l \exp(i\mathbf{K}\cdot\mathbf{R}_{lb}), \tag{2.62}$$

where

$$W_b = -\tfrac{1}{2}\sum_{\mathbf{q}p}[\mathbf{K}\cdot(\mathbf{r}_{lb})_{\mathbf{q}p}]^2. \tag{2.63}$$

$A_{s,e}$ vanishes unless $\mathbf{K}$ is a reciprocal-lattice vector, just as A_s vanishes for the nonvibrating lattice. The factor introduced by the lattice vibration, $\exp(-W_b)$, involves operators $(\mathbf{r}_{lb})^2$. Only the diagonal matrix elements of these operators are nonzero. It follows that $A_{s,e}$ refers to the scattering that does not involve a change of the vibrational state of the lattice, which is called elastic scattering. The photon energy of the frequency of the scattered radiation is the same as that of the incident radiation. The effect of ion vibration in disturbing the coherence leads to a reduction of the scattering amplitude. The intensity of the scattered radiation is proportional to the expectation value of the squared modulus of $A_{s,e}$. The reduction factor of the intensity is known as the Debye–Waller factor.

For a lattice with one ion per unit cell, we have simply the Debye–Waller factor $\langle|\exp(-2W)|\rangle$. With

$$\langle|\exp(-2W)|\rangle \sim \exp[-\langle|2W|\rangle], \tag{2.64}$$

the problem reduces to obtaining $\langle|2W|\rangle$. Usually, only the three acoustical phonon branches are significant. For a given $\mathbf{K}$, we have on the average

$$[\mathbf{K}\cdot\mathbf{r}_{\mathbf{q}p}]^2 = \tfrac{1}{3}K^2(\mathbf{r}_{\mathbf{q}p})^2.$$

It follows that

$$\langle|2W|\rangle = \left\langle\left|\sum_{\mathbf{q}p}[\mathbf{K}\cdot\mathbf{r}_{\mathbf{q}p}]^2\right|\right\rangle = \frac{K^2}{3}\sum_{\mathbf{q}p}\langle|(\mathbf{r}_{\mathbf{q}p})^2|\rangle.$$

Using (2.34), we get

$$\langle|2W|\rangle = \frac{K^2}{3}\sum_{\mathbf{q}p}\frac{\hbar}{NM\omega_{\mathbf{q}p}}(n_{\mathbf{q}p}+\tfrac{1}{2}). \tag{2.65}$$

For the thermal vibration of a lattice, the excitation $n_{\mathbf{q}p}$ of a vibration mode depends only on the temperature and the frequency $\omega_{\mathbf{q}p}$ of the mode. Therefore the summation can be replaced by a integration with respect to frequency:

$$\langle|2W|\rangle = \frac{K^2}{3}\frac{\hbar}{NM}\int_0^{\omega_m}\left[\sum_p \rho_p(\omega)\right]\frac{1}{\omega}(n_\omega+\tfrac{1}{2})\,d\omega.$$

With approximation (2.51) in the Debye theory of specific heat, we get

$$\langle|2W|\rangle = \frac{K^2}{3}\frac{\hbar}{NM}\frac{3}{2\pi^2c^3}\int_0^{\omega_m}\left(\frac{1}{\exp(\hbar\omega/kT)-1}+\frac{1}{2}\right)\omega\,d\omega$$

$$= K^2\frac{\hbar}{NM}\frac{1}{2\pi^2c^3}\left(\frac{kT}{\hbar}\right)^2\int_0^{\hbar\omega_m/kT}\left(\frac{1}{e^x-1}+\frac{1}{2}\right)x\,dx$$

$$= 3\frac{(\hbar K)^2}{M}\frac{1}{k\Theta}\left(\frac{T}{\Theta}\right)^2\int_0^{\Theta/T}\left(\frac{1}{e^x-1}+\frac{1}{2}\right)x\,dx. \quad (2.66)$$

At high temperatures, $\Theta/T \ll 1$,

$$\langle|2W|\rangle = 3\frac{\hbar^2K^2/M}{k\Theta}\left(\frac{T}{\Theta}\right). \quad (2.67a)$$

At low temperatures, $\Theta/T \gg 1$,

$$\langle|2W|\rangle = \frac{3}{4}\frac{\hbar^2K^2/M}{k\Theta}. \quad (2.67b)$$

b. Inelastic Scattering

Consider the second term in (2.60):

$$\sum_{\mathbf{q}p} i\mathbf{K}\cdot(\mathbf{r}_{lb})_{\mathbf{q}p}\prod_{\mathbf{q}'p'\neq\mathbf{q}p}\{1-\tfrac{1}{2}[\mathbf{K}\cdot(\mathbf{r}_{lb})_{\mathbf{q}'p'}]^2\} = \sum_{\mathbf{q}p} i\mathbf{K}\cdot(\mathbf{r}_{lb})_{\mathbf{q}p}[1+\cdots]. \quad (2.68)$$

The second and succeeding terms in the square brackets give to this expression terms of the third and higher orders in r. They will be neglected. The amplitude of scattering contributed by the second term of (2.60) is then

$$A_{s,i} \propto A\exp(i\mathbf{k}'\cdot\mathbf{R})\sum_{\mathbf{q}p}\sum_{lb}(\hat{e}_s\alpha_{b\mathbf{q}p}\hat{e})\exp(i\mathbf{k}\cdot\mathbf{R}_{lb})\mathbf{K}\cdot(\mathbf{r}_{lb})\mathbf{q}p.$$

In view of (2.33), we get

$$A_{s,i} \propto A \exp(i\mathbf{k}' \cdot \mathbf{R}) \left\{ \sum_b F_b \exp(i\mathbf{K} \cdot \mathbf{d}_b) \sum_{\mathbf{q}p} (\hat{e}_s \alpha_{b\mathbf{q}p} \hat{e}) \left(\frac{\hbar}{2NM_b \omega_{\mathbf{q}p}} \right)^{1/2} \right.$$
$$\times \left(\mathbf{e}^*_{b\mathbf{q}p} a_{\mathbf{q}p} \sum_l \exp[i(\mathbf{K} - \mathbf{q}) \cdot \mathbf{L}] \right.$$
$$\left. \left. + \mathbf{e}_{b\mathbf{q}p} a^+_{\mathbf{q}p} \sum_l \exp[i(\mathbf{K} + \mathbf{q}) \cdot \mathbf{L}] \right) \right\} \cdot \mathbf{K}, \qquad (2.69)$$

where $\mathbf{d}_b = \mathbf{R}_{lb} - \mathbf{L}$. It has been pointed out in connection with (2.33) that $a_{\mathbf{q}p}$ and $a^+_{\mathbf{q}p}$ have no diagonal matrix elements and that $\mathbf{r}_{lb}$ has zero expectation value. Therefore $A_{s,i}$ does not occur without a change of lattice vibration in the course of scattering. The term with $a^+_{\mathbf{q}p}$ in the curly brackets may contribute if $n_{\mathbf{q}p}$ changes to $n_{\mathbf{q}p} - 1$, while the term with $a^*_{\mathbf{q}p}$ may be effective for a change from $n_{\mathbf{q}p}$ to $n_{\mathbf{q}p} + 1$. The two cases involve a transition of the vibration state, with the absorption and emission, respectively, of a phonon. Conservation of energy is required for a transition to occur:

$$\Delta(\text{vibration energy}) + \Delta(\text{radiation energy}) = 0. \qquad (2.70)$$

Furthermore, each term in the curly brackets contains a sum that imposes a condition on $\mathbf{K}$ and $\mathbf{q}$ for them not to vanish. Together, the conditions of wavevector conservation and energy conservation are

$$\mathbf{K} - \mathbf{q} = \mathbf{G} \quad \text{and} \quad \hbar\omega_{\mathbf{q}p} = \hbar\omega' - \hbar\omega \quad \text{for phonon absorption,}$$

and

$$\mathbf{K} + \mathbf{q} = \mathbf{G} \quad \text{and} \quad -\hbar\omega_{\mathbf{q}p} = \hbar\omega' - \hbar\omega \quad \text{for phonon emission.} \qquad (2.71)$$

$\hbar\omega$ is the energy quantum of the incident wave, and $\hbar\omega'$ is that of the scattered wave. The wavevector conservation expresses momentum conservation, $\hbar\mathbf{K}$ or $\hbar\mathbf{q}$ being the momentum of a wave packet with wavevectors in the vicinity of $\mathbf{K}$ or $\mathbf{q}$. The scattering $A_{s,i}$ involving a change in the energy quantum of the wave is inelastic scattering.

$\omega' - \omega$ is called the frequency shift. Waves with $\omega' < \omega$ are called the Stokes component, and those with $\omega' > \omega$ are the anti-Stokes component. The scattering that involves acoustic phonons is known as Brillouin scattering, and the effect that involves phonons of optical branches is known as Raman scattering. The latter has frequency shifts of much larger magnitude.

The parameter α in expression (2.69) for $A_{s,i}$ is called the polarization tensor. Since each term of $A_{s,i}$ involves either emission or absorption of a phonon, α depends on the point-group symmetry of the relevant phonon in the structure.

Background scattering. Although a change of lattice vibration is necessary for $A_{s,i}$ to appear, it is not necessary for the scattering intensity, which is proportional to $A_{s,i}A_{s,i}^*$. In view of (2.29),

$$\left\{\mathbf{e}_{b\mathbf{q}p}^* a_{\mathbf{q}p}\sum_l \exp[i(\mathbf{K}-\mathbf{q})\cdot\mathbf{L}] + \mathbf{e}_{b\mathbf{q}p}a_{\mathbf{q}p}^+\sum_l \exp[i(\mathbf{K}+\mathbf{q})\cdot\mathbf{L}]\right\}^*$$
$$= \mathbf{e}_{b\mathbf{q}p}a_{\mathbf{q}p}^+\sum_l \exp[-i(\mathbf{K}-\mathbf{q})\cdot\mathbf{L}]$$
$$+ \mathbf{e}_{b\mathbf{q}p}^* a_{\mathbf{q}p}\sum_l \exp[-i(\mathbf{K}+\mathbf{q})\cdot\mathbf{L}].$$

Therefore $A_{s,i}A_{s,i}^*$ contains terms of the following kind:

$$\mathbf{e}_{b\mathbf{q}p}^2 a_{\mathbf{q}p}a_{\mathbf{q}p}^+\left(\sum_l \exp[i(\mathbf{K}-\mathbf{q})\cdot\mathbf{L}]\right)\left(\sum_l \exp[-i(\mathbf{K}-\mathbf{q})\cdot\mathbf{L}]\right)$$
$$+ e_{b\mathbf{q}p}^2 a_{\mathbf{q}p}^+ a_{\mathbf{q}p}\left(\sum_l \exp[i(\mathbf{K}+\mathbf{q})\cdot\mathbf{L}]\right)\left(\sum_l \exp[-i(\mathbf{K}+\mathbf{q})\cdot\mathbf{L}]\right). \quad (2.72)$$

Since $a_{\mathbf{q}p}a_{\mathbf{q}p}^+$ and $a_{\mathbf{q}p}^+a_{\mathbf{q}p}$ have nonvanishing expectation values, these terms contribute to the intensity of scattering of unchanged frequency. The contribution is subject to

$$\mathbf{K}-\mathbf{q}=\mathbf{G} \qquad \text{for } \mathbf{a}_{\mathbf{q}p}a_{\mathbf{q}p}^+ \quad (2.73)$$

and

$$\mathbf{K}+\mathbf{q}=\mathbf{G} \qquad \text{for } a_{\mathbf{q}p}^+a_{\mathbf{q}p}.$$

For a primary radiation of wavevector $\mathbf{k}_0$, any $\mathbf{k}'=\mathbf{K}-\mathbf{k}_0$ or any angle of scattering can satisfy each of the conditions with a suitable $\mathbf{q}$. In this sense, the effect may be considered background scattering.

c. Scattering of Slow Neutrons

Inelastic scattering provides an effective means for determining the dispersion relation $\omega_{\mathbf{q}}(\mathbf{q})$ of the phonons. An x ray has a wavevector of magnitude $\leq 10^7\ \text{cm}^{-1}$; radiations of longer wavelengths have proportionately smaller k's. On the other hand, the smallest magnitude of a reciprocal-lattice vector is $2\pi/a \sim 10^8$ cm^{-1}, and the maximum magnitude of $\mathbf{q}$ is one-half of this value. Therefore the conditions of wavevector conservation in (2.71) must have $\mathbf{G} = 0$, giving

$$\mathbf{q} = \pm\mathbf{K}.$$

Thus, the scattering of electromagnetic radiation covers only a small range of $\mathbf{q}$.

For neutron scattering, the ratio of scattered to incident amplitude is similar to that of photon scattering, except that neutron spin is usually thought of instead of radiation polarization. The conditions for inelastic scattering, Eq. (2.71), remain the same. The kinetic energy E of a neutron is related to the wavevector $\mathbf{k}$ according to

$$E = \hbar^2 k^2 / 2M,$$

M being the mass of a neutron. It follows that

$$k = 2.2 \times 10^9 E^{1/2} \text{cm}^{-1}, \tag{2.74}$$

where E is in units of electron volts. Slow neutrons with E of the order 10^{-2} eV have k's comparable to the magnitude of the smallest reciprocal-lattice vector. It is easily seen that the wavevector conservation conditions can be satisfied for the whole range of $\mathbf{q}$ in the inelastic scattering of such neutrons. With regard to the determination of phonon energy, $\hbar\omega_{\mathbf{q}} = |\Delta E|$ according to the energy conservation of inelastic scattering. It is desirable for the accuracy of the determination to have $|\Delta E|/E = \hbar\omega_{\mathbf{q}}/E$ as large as possible. Although a given $\mathbf{q}$ may be covered by using neutrons of various E's, the smallest E is preferred.

d. The Mössbauer Effect

Gamma rays emitted by a radioactive nucleus may have an extremely small natural linewidth, for example, $\sim 10^{-10}$ eV for zinc-67. They provide a precise measure of the separation between the nuclear levels involved in the transition, thereby serving to investigate the environment of the nucleus through its effect on the nuclear levels. The following questions should be considered.

The energy $\hbar\omega$ of the gamma ray emitted by a nucleus at rest differs from the energy separation E of the nuclear levels due to the recoil of the nucleus. The speed of the recoiling nucleus, v, can be obtained from the principle of conservation of momentum:

$$Mv = \hbar\omega/c,$$

where M is the mass of the nucleus. The recoil energy of the nucleus,

$$R = \hbar^2K^2/2M = Mv^2/2 = (\hbar\omega)^2/2Mc^2, \tag{2.75}$$

gives the frequency shift $\Delta\omega = -R/\hbar$ of the γ ray due to the recoil of the nucleus. The order of magnitude of the frequency shift,

$$\Delta\omega \sim 10^{15}\ \mathrm{s}^{-1}, \tag{2.76}$$

is much larger than the natural linewidth $10^8\ \mathrm{s}^{-1}$ of the γ ray.

The second question to be considered is the Doppler effect. A γ ray of frequency ω emitted by a nucleus moving with a speed v toward the observer appears to have a frequency $\omega(1+v/c)$ instead of ω. However, a nucleus vibrating about a fixed equilibrium position in a lattice has $\langle|v|\rangle = \langle|p|\rangle/m = 0$, and therefore the frequency of the emitted γ ray is not broadened by the Doppler effect. A similar consideration applies also to the phenomenon of light scattering by a lattice with ion vibration.

Emission involves considerations analogous to those of scattering. The product of the incident radiation A and the form factor F of the scatterer is replaced by a factor characteristic of the emitting nucleus. Elastic scattering from a lattice with vibration corresponds to the recoilless emission from a nucleus that is one of the vibrating ions in the lattice; recoilless emission is known as the Mössbauer effect. Inelastic scattering corresponds to emission with recoil of the nucleus.

Recoilless emission has the frequency of the emission from a stationary nucleus and a narrow width of the order of the natural linewidth. The emission intensity is reduced by the vibration of the nucleus, being proportional to the same Debye–Waller factor as the intensity of elastic scattering. For a lattice at low temperatures, $T \ll \Theta$,

$$\langle|2W|\rangle = \frac{3}{2}\frac{\hbar^2K^2/2M}{k\Theta} = \frac{3}{2}\frac{R}{k\Theta}, \tag{2.77}$$

according to (2.67b) and (2.75). As to the emission with recoil of the nucleus, the frequency is lowered and greatly broadened, the process being associated with the emission of one of various phonons.

6. THERMAL EXPANSION

Up to now, lattice vibration has been treated in the approximation of harmonic motion of the ions. The vibration energy was considered only up to terms of the second order in the displacement of the ions. Higher-order terms produce anharmonicity of vibration, which is basic for thermal expansion and phonon–phonon scattering.

Let α be the coefficient of thermal expansion of volume, for a crystal having volume V_0 at $T = 0$ and $P = 0$:

$$\alpha = \frac{1}{V_0}\frac{dV}{dT}. \tag{2.78}$$

At $T = 0$, the pressure and volume are related through the compressibility χ_0:

$$P = -\frac{V - V_0}{V_0}\frac{1}{\chi_0}$$

and

$$E(V) - E(V_0) + \int_{V_0}^{V} P\,dV = 0. \tag{2.79}$$

E is the internal energy, which depends only on the volume and not directly on the temperature; it includes the zero point energy of lattice vibration. In the expansion

$$E(V) - E(V_0) = (V - V_0)\left(\frac{dE}{dV}\right)_{V_0} + \frac{1}{2}(V - V_0)^2\left(\frac{d^2E}{dV^2}\right)_{V_0} + \cdots, \tag{2.80}$$

the first term is zero as required by the possible existence of equilibrium at $T = 0$ and $P = 0$. The integral on the left hand side of the above equation is

$$\int_{V_0}^{V} P\,dV = -\frac{1}{2}\frac{(V - V_0)^2}{\chi_0 V_0}.$$

Therefore,

$$\frac{1}{\chi_0} = V_0\left(\frac{d^2E}{dV^2}\right)_{V_0}. \tag{2.81}$$

Consider now the expansion

$$\frac{dE}{dV} = \left(\frac{dE}{dV}\right)_{V_0} + (V - V_0)\left(\frac{d^2E}{dV^2}\right)_{V_0} + \cdots,$$

the first term of which is zero. Hence we get

$$\frac{dE}{dV}=\frac{1}{\chi_0}\frac{V-V_0}{V_0}. \tag{2.82}$$

Besides E, the Helmholtz free energy F contains a contribution from the lattice vibration that depends on T. The contribution F_v is easily shown to be

$$F_v = kT\sum_i \ln[1-\exp(-\hbar\omega_i/kT)], \tag{2.83}$$

where i is the index of the phonon mode. The condition for equilibrium under $P=0$ gives

$$0=\left(\frac{\partial F}{\partial V}\right)_T=\frac{dE}{dV}+\left(\frac{\partial F_v}{\partial V}\right)_T. \tag{2.84}$$

It follows that

$$\frac{1}{\chi_0}\frac{V-V_0}{V_0}+\sum_i \frac{\hbar\omega_i}{\exp(\hbar\omega_i/kT)-1}\frac{d(\ln\omega_i)}{dV}=0. \tag{2.85}$$

Make the approximation that $d(\ln\omega_i)/dV$ is the same for all the modes. Differentiate the equation with respect to T, taking $d(\ln\omega)/dV$ to be a constant. Multiplying the result by χ_0, we get

$$\alpha=\frac{1}{V_0}\frac{dV}{dT}=-\chi_0 C_v\frac{d(\ln\omega)}{dV}=-\chi_0\frac{C_v}{V}\frac{d(\ln\omega)}{d(\ln V)}\simeq-\chi_0\frac{C_v}{V_0}\frac{d(\ln\omega)}{d(\ln V)}, \tag{2.86}$$

where C_v is the specific heat at constant volume. The factor

$$\gamma=\frac{\alpha V_0}{\chi_0 C_v}=-\frac{d(\ln\omega)}{d(\ln V)} \tag{2.87}$$

is known as the Grüneisen constant. It is independent of T in this approximation. The conclusion is verified over some wide temperature range for a number of solids.

The factor $d(\ln\omega)/d(\ln V)$ basic for thermal expansion originates from anharmonicity of lattice vibration. The energy E_R in the equation of motion of the lattice may be expanded in terms of the ion displacements:

$$E_R = E_0 + \frac{1}{2} \sum_{lb,l'b'} G_{lb,l'b'} \mathbf{r}_{lb} \mathbf{r}_{l'b'} + \frac{1}{6} \sum_{lb,l'b',l''b''} B_{lb,l'b',l''b''} \mathbf{r}_{lb} \mathbf{r}_{l'b'} \mathbf{r}_{l''b''} + \cdots . \tag{2.88}$$

Expression (2.6), used for the treatment of lattice vibration, is limited to the first two terms, and the approximation gives harmonic vibration. Consider now the effect of including the third term. Let the lattice be expanded by a factor $1+\epsilon$ in volume. For simplicity, consider the expansion to be uniform in direction. Let $\mathbf{r}'$ be the displacement from the equilibrium of an ion in the expanded lattice:

$$\mathbf{r}_{lb} = \frac{1}{3}\epsilon(\mathbf{L}+\mathbf{b}) + \mathbf{r}'_{lb}, \tag{2.89}$$

where $\mathbf{b}$ is the vector distance from the origin of cell l to ion b. Substituting this expression of $\mathbf{r}_{lb}$'s, we get

$$E_R = (E_0 + C\epsilon^2) + \frac{1}{2} \sum_{lb,l'b'} G'_{lb,l'b'} r'_{lb} r'_{l'b'} + \cdots . \tag{2.90}$$

$C\epsilon^2$ comes from the terms with $G_{lb,l'b'}$; it represents the change in elastic energy due to expansion. Terms linear in $\mathbf{r}'$ vanish, since $\mathbf{r}'$ is displacement from the equilibrium position of the ion in the expanded lattice. In the terms of second order in $\mathbf{r}'$,

$$G'_{lb,l'b'} = G_{lb,l'b'} + \frac{1}{3} \sum_{l''b''} B_{lb,l'b',l''b''} (\mathbf{b}'' + \mathbf{L}''). \tag{2.91}$$

Since G is changed to G' on account of the terms with B, the frequencies of vibration in the harmonic approximation are changed by the volume expansion.

7. LATTICE THERMAL CONDUCTIVITY

Normal modes of lattice vibration, phonons, are extended waves. Transport phenomena involve the concept of wave packets, which have limited dimensions in space. The spatial extension $\delta\mathbf{r}$ of a wave packet is related to the spread of wave numbers, $\delta\mathbf{q}$, of the constituent waves by the uncertainty principle:

$$\delta x \, \delta q_x \sim 1.$$

The uncertainty in energy δE associated with the spread δq is

$$\delta E = \hbar(\partial\omega/\partial q)\,\delta q. \tag{2.92}$$

An energy resolution of ΔE corresponds to

$$\delta x = \frac{1}{\delta q_x} = \frac{\hbar(\partial\omega/\partial q_x)}{\Delta E}. \tag{2.93}$$

For thermal properties, we are interested in having $\Delta E < kT$. Therefore the wave packets of interest have dimensions

$$\delta x \geq \frac{\hbar(\partial\omega/\partial q_x)}{kT}. \tag{2.94}$$

For acoustical phonons, which are of primary interest, the group velocity $\partial\omega/\partial q$ is the velocity of sound c, and $\hbar c = k\Theta/(6\pi^2 N)^{1/3} \sim k\Theta a$, where a is the lattice constant. The condition

$$\delta x \geq (\Theta/T)a$$

is reasonable.

According to the kinetic theory, the thermal conductivity of particles, wave packets, is given by the generalized expression

$$\kappa = \tfrac{1}{3}C\bar{v}\lambda, \tag{2.95}$$

where C is the heat capacity per unit volume, $\bar{v}$ is the average thermal velocity, and λ is the mean free path of the particles. For a concrete treatment, we have to resort to the Boltzmann equation of transport. Consider $f_{\mathbf{q}p}$, the statistical average value of the quantum number $n_{\mathbf{q}p}$ of mode $\mathbf{q}p$. It may be regarded as the number of phonons of the mode $\mathbf{q}p$. For the present problem, $f_{\mathbf{q}p}$ is the number of wave packets centered around $\mathbf{q}p$. It may be referred to as the distribution function of phonon particles. The variation of $f_{\mathbf{q}p}$ with time produced by the drift motion of the particles is

$$(\partial f_{\mathbf{q}p}/\partial t)_d = -\nabla_{\mathbf{q}}\omega_{\mathbf{q}p}\cdot\nabla_{\mathbf{r}} f_{\mathbf{q}p}, \tag{2.96}$$

where $\nabla_{\mathbf{q}}\omega_{\mathbf{q}p}$ is the velocity of the wave packet $\mathbf{q}p$. The scattering of phonon particles also causes $f_{\mathbf{q}p}$ to vary with time; this variation will be denoted by $(\partial f_{\mathbf{q}p}/\partial t)_s$. The Boltzmann equation is

$$\frac{df_{\mathbf{q}p}}{dt} = \left(\frac{\partial f_{\mathbf{q}p}}{\partial t}\right)_d + \left(\frac{\partial f_{\mathbf{q}p}}{\partial t}\right)_s, \tag{2.97}$$

in which $df_{\mathbf{q}p}/dt = 0$ for the steady state. The equation serves to determine the distribution function $f_{\mathbf{q}p}$ for the calculation of the desired property. Thermal conductivity deals with the flux density of thermal energy, which is given by

$$\mathbf{h} = \sum_p \int f_{\mathbf{q}p} \hbar\omega_{\mathbf{q}p} \nabla_{\mathbf{q}} \omega_{\mathbf{q}p}\, d\mathbf{q}. \tag{2.98}$$

For a lattice in thermal equilibrium at a uniform temperature T, the distribution function is given by

$$f^0_{\mathbf{q}p}(T) = \frac{1}{\exp(\hbar\omega_{\mathbf{q}p}/kT) - 1}. \tag{2.99}$$

In this case, df/dt, $(\partial f/\partial t)_d$, $(\partial f/\partial t)_s$, and the heat flux are obviously each equal to zero. For the steady state, $df_{\mathbf{q}p}/dt = 0$, of a lattice with nonuniform temperature, $T(\mathbf{r}) \neq$ constant, the distribution function may be expressed as

$$f_{\mathbf{q}p}(\mathbf{r}) = f^0_{\mathbf{q}p}(T(\mathbf{r})) + \Delta f_{\mathbf{q}p}(\mathbf{r}). \tag{2.100}$$

For $(\partial f_{\mathbf{q}p}/\partial t)_d$, $\Delta f_{\mathbf{q}p}$ may be neglected; it is then straightforward to obtain $(\partial f_{\mathbf{q}p}/\partial t)_d$ for the given $T(\mathbf{r})$. On the other hand, $\Delta f_{\mathbf{q}p}$ cannot be neglected for $(\partial f_{\mathbf{q}p}/\partial t)_s$, which vanishes in the absence of $\Delta f_{\mathbf{q}p}$. The calculation of $(\partial f_{\mathbf{q}p}/\partial t)_s$ requires detailed consideration. There are various types of scattering that affect $\mathbf{q}p$ of the particle: phonon–phonon scattering, scattering by lattice imperfections, scattering by the boundary of lattice, and electron–lattice scattering in the presence of conduction electrons. Each type of scattering requires special consideration. The following is a brief discussion of phonon–phonon scattering, which is inherent to a perfect lattice.

The third term of E_R in expression (2.88) is the leading term for anharmonicity. With reference to (2.33), it is seen to consist of a number of triple products of quantization operators. A creation operator a^+ can increase the quantum number of the mode or the number of phonons in the mode by 1. An annihilation operator has the opposite effect. As a perturbation, anharmonicity produces transitions between two stationary states of harmonic oscillation. A triple product of quantization operators that is effective for producing transitions cannot contain exclusively either creation operators or annihilation operators, on account of energy conservation. An effective triple product $a^+_{\mathbf{q}p} a^+_{\mathbf{q}'p'} a_{\mathbf{q}''p''}$ or $a_{\mathbf{q}p} a_{\mathbf{q}'p'} a^+_{\mathbf{q}''p''}$ enters into anharmonicity in the form

$$\sum_{ll'l''} \exp[i(\mathbf{q}\cdot\mathbf{L} + \mathbf{q}'\cdot\mathbf{L}' - \mathbf{q}''\cdot\mathbf{L}'')] a^+_{\mathbf{q}p} a^+_{\mathbf{q}'p'} a_{\mathbf{q}''p''} \tag{2.101}$$

or

$$\sum_{ll'l''} \exp[i(-\mathbf{q}\cdot\mathbf{L} - \mathbf{q}'\cdot\mathbf{L}' + \mathbf{q}''\cdot\mathbf{L}'')]a_{\mathbf{q}p}a_{\mathbf{q}'p'}a^{+}_{\mathbf{q}''p''}.$$

These quantities should not be affected if every $\mathbf{L}$ is changed by the same lattice vector, since such a change amounts to merely taking a different lattice point as origin. Therefore, these quantities must vanish unless

$$\mathbf{q} + \mathbf{q}' - \mathbf{q}'' = \mathbf{G}, \tag{2.102}$$

where $\mathbf{G}$ is a reciprocal-lattice vector. $\mathbf{q}$ and $\mathbf{q}'$ refer to the two normal modes that are associated with the same kind of operators, creation or annihilation. Lattice transitions with $\mathbf{G} = 0$ are called normal processes, or N processes, and those with $\mathbf{G} \neq 0$ are known as umklapp processes, or U processes. The energy-conservation condition required for a transition is

$$\hbar\omega_{\mathbf{q}} + \hbar\omega_{\mathbf{q}'} = \hbar\omega_{\mathbf{q}''}. \tag{2.103}$$

It is important to note that scatterings by N processes do not produce a resistance to heat flow. Therefore they play no role in determining the thermal conductivity. Consider the quantity

$$\mathbf{P} = \sum_{\mathbf{q}p} f_{\mathbf{q}p}\hbar\mathbf{q},$$

which represents the total momentum of the phonon particles. For a lattice in uniform thermal equilibrium, the flow of thermal energy $\mathbf{h}$ and the total momentum $\mathbf{P}$ are both equal to zero. In the case of a temperature variation in the lattice, $\mathbf{h} \neq 0$ and $\mathbf{P} \neq 0$. Since N processes do not affect $\mathbf{P}$, they cannot lead to the equilibrium condition $\mathbf{P} = 0$ and $\mathbf{h} = 0$ when the temperature becomes uniform. In other words, some heat flow persists under uniform temperature equilibrium. A mechanism that does not dissipate heat flow gives no resistance for the heat current.

With $\mathbf{G} \neq 0$, U processes are the phonon–phonon scatterings that produce thermal resistance. Accurate calculations are complicated, requiring detailed information on the phonon spectrum and anharmonicity terms of E_R. Various treatments using reasonable approximations give expressions of the following form for the thermal conductivity exclusively due to U processes of scattering:

$$\begin{aligned} &\kappa \propto \left(\frac{T}{\Theta}\right)^3 \exp\left(\frac{\Theta}{bT}\right) && \text{for} \quad T < \Theta, \\ &\kappa \propto \Theta/T && \text{for} \quad T > \Theta, \end{aligned} \tag{2.104}$$

where Θ is the Debye temperature and $b \approx 2$.

CHAPTER THREE
System of Electrons

1. ONE-ELECTRON APPROXIMATION

In the adiabatic approximation outlined in Section II.1, the function ψ of electronic coordinates is an eigenfunction of the operator

$$H = -\sum_j \frac{\hbar^2}{2m}\nabla_j^2 + \sum_{j,k}\sum{}' \frac{e^2}{2r_{jk}} + V_{\mathrm{ei}}(\mathbf{r}_1,\mathbf{r}_2,\ldots,\mathbf{R}_1,\mathbf{R}_2) + V_{\mathrm{ii}}(\mathbf{R}_1,\mathbf{R}_2,\ldots) \qquad (3.1)$$

for a given set of positions $\mathbf{R}$ of ions in the lattice. It is the accepted approach to consider ψ's for the equilibrium positions of the ions, referring to them as electronic wavefunctions. The effect of ion vibration is treated as perturbation.

The one-electron approximation, approximating ψ by a combination of one-electron functions, is used in most developments of the theory. Two kinds of combinations have been adopted. In the Hartree approximation, ψ is assumed to be simply a product of one-electron functions φ, with a different function for each electron in view of the Pauli exclusion principle. The variational theorem

$$\delta\int \psi^* H\psi = 0$$

leads, then, to the following Hamiltonian operator for the one-electron functions:

$$H_i^{\mathrm{H}} = -\frac{\hbar^2}{2m}\nabla_i^2 + V(\mathbf{r}_i) + \sum_j{}' e^2 \int \frac{|\varphi_j|^2}{r_{ij}} d\tau_j \tag{3.2}$$

where i and j are electron indices. Since the last term makes the Hamiltonian for one electron depend on the functions of all the other electrons, a self-consistent procedure is used: Assume a set of functions for all the electrons, check the consistency of the solutions with the assumed functions, and iterate corrections of the assumed functions until an acceptable consistency is obtained.

Electrons, being fermions, should have a wavefunction antisymmetric under permutation of the electrons, according to the Pauli exclusion principle. The better combination of one-electron functions is the Slater determinant, which is antisymmetric in contrast to a simple product. Each one-electron function of $\mathbf{r}$ corresponds to one or the other eigenvalue, $+\hbar/2$ or $-\hbar/2$, of a spin component. Functions corresponding to different values of s_z are orthogonal, and different functions corresponding to the same s_z are orthogonal to each other. The determinant is

$$\psi = \frac{1}{\sqrt{n!}} \begin{vmatrix} \varphi_\alpha(1) & \varphi_\alpha(2) & \cdots & \varphi_\alpha(n) \\ \varphi_\beta(1) & \varphi_\beta(2) & \cdots & \varphi_\beta(n) \\ & & \vdots & \\ \varphi_\nu(1) & \varphi_\nu(2) & \cdots & \varphi_\nu(n) \end{vmatrix}, \tag{3.3}$$

where $1, 2, \ldots, n$ are indices of electrons and $\alpha, \beta, \ldots, \nu$ are indices of the different functions. The variational theorem leads to a set of equations of the following type for the one-electron functions:

$$H^{\mathrm{F}} \varphi_\alpha(i) = \epsilon_\alpha \varphi_\alpha(i). \tag{3.4}$$

The Fock Hamiltonian operator H^{F} stands for

$$H^{\mathrm{F}} = -\frac{\hbar^2}{2m}\nabla_i^2 + V(\mathbf{r}_i) + \sum_\beta{}' e^2 \int \frac{|\varphi_\beta(j)|^2}{r_{ij}} d\tau_j + A, \tag{3.5}$$

where

$$A\varphi_\alpha(i) = -\sum_\beta{}' e^2 \varphi_\beta(i) \int \frac{\varphi_\beta^*(j)\varphi_\alpha(j)}{r_{ij}} d\tau_j$$

$$= \left\{ -e^2 \sum_\beta{}' \left[\int d\tau_j \frac{\varphi_\beta^*(j)\varphi_\alpha(j)\varphi_\beta(i)\varphi_\alpha^*(i)}{r_{ij}} \right] \times \frac{1}{\varphi_\alpha(i)\varphi_\alpha^*(i)} \right\} \varphi_\alpha(i).$$

Equation (3.4) is often referred to as the Hartree–Fock approximation equation.

In comparison with H^{H} of the Hartree approximation, the Fock Hamiltonian H^{F} has an additional operator A that results from the use of a determinantal wavefunction instead of a product wavefunction. This operator has the effect of correlating the electronic motions. The expectation value of A is called the exchange energy. The expression of $A\varphi_\alpha(i)$ shows that A affects only the correlation of electrons of parallel spin, not the correlation of electrons with antiparallel spins. Clearly, the introduction of exchange does not fully account for the effect of correlation. As a simple example, consider the system of N free electrons in a background of uniformly distributed positive charge of total magnitude Ne. The exchange energy is calculated to be

$$-0.458e^2/r_s,$$

where

$$(4\pi/3)r_s^3 = 1/n,$$

n being the number of electrons per unit volume. For sufficiently large r_s, the electrons form a body-centered cubic arrangement, and the energy is changed by

$$-0.746e^2/r_s$$

relative to that in the absence of correlation. The exchange energy is short of the correlation energy by

$$-0.746e^2/r_s - (-0.458e^2/r_s) = -0.288e^2/r_s.$$

The difficulty of properly treating the correlation is basic to the use of the one-electron approximation for a system of many electrons. The approach has been to start with the one-electron approximation and use some additional refinement to deal with a specific aspect of collective behavior.

Koopmans' theorem. The total energy E_N of a system of large number N of electrons is

$$E_N = \sum_i^N \epsilon_i - \tfrac{1}{2}\sum_i^N (\epsilon_c)_i, \tag{3.6}$$

where ϵ_c is the correlation energy contained in the one-electron energy ϵ. The second term on the right-hand side prevents double counting of the interaction between a pair of electrons, once in the ϵ of the first electron and once in the ϵ of the second. It can be shown that for a system of large N such that the remaining φ's and ϵ's are not appreciably affected by the removal of an electron from a state j,

$$E_N - E_{N-1} = \epsilon_j. \tag{3.7a}$$

Furthermore, the change in energy due to shifting an electron from state α to state β is given by

$$E(\beta) - E(\alpha) = \epsilon_\beta - \epsilon_\alpha. \tag{3.7b}$$

These conclusions are obviously very important for the physical significance of the one-electron approximation. Deduced first by Koopmans, they are known as Koopmans' theorem.

2. CONCEPT OF ELECTRONIC ENERGY BANDS

Electronic energy bands are based on the one-electron approximation. Write the Hartree–Fock equation in the form

$$\left[-\frac{\hbar^2}{2m}\nabla_{\mathbf{r}}^2 + V(\mathbf{r})\right]\varphi(\mathbf{r}) = \epsilon\varphi(\mathbf{r}), \tag{3.8}$$

where V stands for the last three terms on the right-hand side of (3.5) and only the variables $\mathbf{r}$ are indicated for V and φ. Changing the origin of spatial coordinates by $\mathbf{R}$ makes

$$\mathbf{r} = \mathbf{r}' + \mathbf{R},$$

where $\mathbf{r}'$ refers to the new origin. Substituting $\mathbf{r}' + \mathbf{R}$ for $\mathbf{r}$, we get

$$\left[-\frac{\hbar^2}{2m}\nabla_{\mathbf{r}'}^2 + V(\mathbf{r}' + \mathbf{R})\right]\varphi(\mathbf{r}' + \mathbf{R}) = \epsilon\varphi(\mathbf{r}' + \mathbf{R}).$$

Consider now $\mathbf{R}=\mathbf{a}_n$, where $\mathbf{a}_n$ is a lattice vector of the crystal being considered. Since a crystal has translational symmetry, we have

$$V(\mathbf{r}'+\mathbf{a}_n)=V(\mathbf{r}'). \tag{3.9}$$

Therefore,

$$\left[-\frac{2}{2m}\nabla^2_{\mathbf{r}'}+V(\mathbf{r}')\right]\varphi(\mathbf{r}'+\mathbf{a}_n)=\epsilon\varphi(\mathbf{r}'+\mathbf{a}_n).$$

On the other hand, the Hartree–Fock equation in terms of $\mathbf{r}'$ is

$$\left[-\frac{\hbar^2}{2m}\nabla^2_{\mathbf{r}'}+V(\mathbf{r}')\right]\varphi(\mathbf{r}')=\epsilon\varphi(\mathbf{r}').$$

It follows that, on account of the translational symmetry of a crystal, $\varphi(\mathbf{r})$ and $\varphi(\mathbf{r}+\mathbf{a}_n)$ are eigenfunctions of the same eigenvalue of the Fock Hamiltonian.

a. Consider first the case where the solution $\varphi(\mathbf{r})$ is nondegenerate. Then $\varphi(\mathbf{r}+\mathbf{a}_n)$ must be $\varphi(\mathbf{r})$ multiplied by a constant:

$$\varphi(\mathbf{r}+\mathbf{a}_n)=C(\mathbf{a}_n)\varphi(\mathbf{r}). \tag{3.10}$$

For $\varphi(\mathbf{r})$ and $\varphi(\mathbf{r}+\mathbf{a}_n)$ both normalized,

$$|C(\mathbf{a}_n)|^2=1.$$

Since

$$C(\mathbf{a}_m+\mathbf{a}_n)\varphi(\mathbf{r})=C(\mathbf{a}_m)[C(\mathbf{a}_n)\varphi(\mathbf{r})]=C(\mathbf{a}_m)C(\mathbf{a}_n)\varphi(\mathbf{r}),$$

we have

$$C(\mathbf{a}_m)C(\mathbf{a}_n)=C(\mathbf{a}_m+\mathbf{a}_n).$$

These properties of the C's lead to

$$C(\mathbf{a}_n)=\exp(i\mathbf{k}\cdot\mathbf{a}_n). \tag{3.11}$$

$\mathbf{k}$ is called the wavevector. A consideration similar to that discussed for the wavevector of lattice waves shows that only reduced wavevectors need be considered; that is, $\mathbf{k}$ may be limited to the first Brillouin zone. Furthermore, it follows from the Born–von Kármán boundary condition that $\mathbf{k}$ is quasi-continuous.

Let $\varphi(\mathbf{r})$ associated with a wavevector $\mathbf{k}$ be expressed in the form

$$\varphi_{\mathbf{k}}(\mathbf{r})=\exp(i\mathbf{k}\cdot\mathbf{r})u_{\mathbf{k}}(\mathbf{r}). \tag{3.12}$$

Then

$$\varphi_{\mathbf{k}}(\mathbf{r}+\mathbf{a}_n) = \exp(i\mathbf{k}\cdot\mathbf{a}_n)\exp(i\mathbf{k}\cdot\mathbf{r})u_{\mathbf{k}}(\mathbf{r}+\mathbf{a}_n).$$

We have seen that

$$\varphi_{\mathbf{k}}(\mathbf{r}+\mathbf{a}_n) = \exp(i\mathbf{k}\cdot\mathbf{a}_n)\varphi_{\mathbf{k}}(\mathbf{r}) = \exp(i\mathbf{k}\cdot\mathbf{a}_n)\exp(i\mathbf{k}\cdot\mathbf{r})u_{\mathbf{k}}(\mathbf{r})$$

follows from the translationary symmetry. Therefore,

$$u_{\mathbf{k}}(\mathbf{r}+\mathbf{a}_n) = u_{\mathbf{k}}(\mathbf{r}). \tag{3.13}$$

Expression (3.12) for a one-electron function, in which $u_{\mathbf{k}}(\mathbf{r})$ has the periodicity of the lattice, is the important and well-known Bloch function.

Substituting the Bloch function into the Hartree–Fock equation, we get an equation for $u_{\mathbf{k}}(\mathbf{r})$:

$$\left[-\frac{\hbar^2}{2m}(\nabla - i\mathbf{k})^2 + V(\mathbf{r})\right]u_{\mathbf{k}}(\mathbf{r}) = \epsilon_{\mathbf{k}}u_{\mathbf{k}}(\mathbf{r}). \tag{3.14}$$

We indicate the various solutions of $u_{\mathbf{k}}(\mathbf{r})$ and $\epsilon_{\mathbf{k}}$ given by the equation by an additional index in the subscript. Considering the complex conjugate equation, we get

$$u_{-\mathbf{k},m}(\mathbf{r}) = u^*_{\mathbf{k},m}(\mathbf{r}) \qquad \text{and} \qquad \epsilon_m(-\mathbf{k}) = \epsilon_m(\mathbf{k}). \tag{3.15}$$

The index m is the index of the energy band, which is the energy span covered by various $\mathbf{k}$'s for a fixed m. Two adjacent energy bands may be either overlapping or separated by an energy gap ϵ_g.

b. Consider now the case of degenerate $\varphi(r)$. Assume that the set of independent, degenerate functions has been converted by the standard procedure to a set of functions that are normalized and mutually orthogonal. For any one of the functions,

$$\varphi_i(\mathbf{r}+\mathbf{a}_n) = \sum_j C_{ji}\varphi_j(\mathbf{r}). \tag{3.16}$$

Since orthogonality and normalization are not affected by a transformation of coordinates, we have

$$\delta_{ji} = \int \varphi_j^*(\mathbf{r}+\mathbf{a}_n)\varphi_i(\mathbf{r}+\mathbf{a}_n)\,d\tau = \sum_\nu C^*_{\nu j}C_{\nu i} = \sum_\nu (C^+)_{j\nu}C_{\nu i}.$$

Mathematically,

$$\delta_{ji} = \sum_\nu (C^{-1})_{j\nu}C_{\nu i},$$

where C^{-1} is the reciprocal of C. We see that $C^{+} = C^{-1}$; that is, C is a unitary matrix. It is known that a unitary matrix C can be transformed into a diagonal unitary matrix C' by using a suitable unitary matrix P: $C' = P^{-1}CP$. Furthermore, a unitary transformation does not affect the orthogonality of either of the two sets of functions. Therefore we can get

$$\varphi_i(\mathbf{r} + \mathbf{a}_n) = C_{ii}\varphi_i(\mathbf{r}), \tag{3.17}$$

as in the nondegenerate case.

Tight-Binding Approximation. The existence of electronic energy bands with a possible energy gap ϵ_g between two neighboring bands can be intuitively visualized from the point of view of the tight-binding approximation. An electron bound to an isolated atom has discrete energy levels. As atoms are brought close to each other, mutual interaction broadens each discrete level into an energy band. An energy gap exists between two bands if the bands originate from discrete levels with an energy difference large compared to broadening. Calculations can be made by using a linear combination of atomic orbitals (LCAO); the atomic orbitals are those of discrete states of individual atoms. Such calculations are justifiable only for the energy bands that evolve from small overlaps of the relevant atomic orbitals.

Metal, Insulator, and Semiconductor. The probability of an electronic state being occupied by an electron is given by the Fermi–Dirac distribution function characterized by a Fermi energy ϵ_{F}. At temperature $T = 0$, the states with $\epsilon < \epsilon_{\mathrm{F}}(0)$ are fully occupied and the states with $\epsilon > \epsilon_{\mathrm{F}}(0)$ are completely empty. A material with $\epsilon_{\mathrm{F}}(0)$ inside an energy band is a metal; a material with $\epsilon_{\mathrm{F}}(0)$ inside an energy gap is an insulator. The band containing $\epsilon_{\mathrm{F}}(0)$ in a metal is called the conduction band. In an insulator, the energy band immediately above the energy gap is called the conduction band, while the one immediately below the energy gap is called the valence band. If the $\epsilon_{\mathrm{F}}(0)$ in the energy gap is close to the conduction or valence band such that the conduction band has appreciable electron occupation or the valence band is appreciably short of full occupation at a reasonable temperature, the material is considered to be a semiconductor. In a semiconductor, either the relevant energy gap is reasonably narrow or the lattice contains a sufficient number of imperfections.

Free-Electron Approximation. The electrons are taken to be free, in the zeroth order of approximation. The Schrödinger equation for free electrons,

$$-\frac{\hbar^2}{2m}\nabla^2\varphi_0(\mathbf{r}) = \epsilon_0\varphi_0(\mathbf{r}), \tag{3.18}$$

gives the plane-wave solutions:

$$\varphi_0(\mathbf{r}) = \Omega^{-1/2}\exp(-i\mathbf{k}\cdot\mathbf{r}) \qquad \text{and} \qquad \epsilon_0 = \hbar^2k^2/2m, \tag{3.19}$$

where Ω is the volume of the crystal. $V(\mathbf{r})$ of the Fock Hamiltonian is treated as a perturbation. Having the translational symmetry of the lattice, $V(\mathbf{r})$ may be expanded into the series

$$V = \sum_G V_G \exp(2\pi i\mathbf{G}\cdot\mathbf{r}), \tag{3.20}$$

where $\mathbf{G}$ is a reciprocal-lattice vector. The first-order perturbation energy is

$$\epsilon_1 = \int V|\varphi_0|^2\,d\tau = \Omega^{-1}\int V\,d\tau = \overline{V}. \tag{3.21}$$

The second-order perturbation energy is given by

$$\epsilon_2 = \sum \frac{|\int \varphi_0\varphi_0'^{*}V\,d\tau|^2}{\epsilon_0 - \epsilon_0'} = \sum_{\mathbf{k}'} \frac{|\int \exp(i(\mathbf{k}-\mathbf{k}')\cdot\mathbf{r})V\,d\tau|^2}{\epsilon_{\mathbf{k}} - \epsilon_{\mathbf{k}'}}$$

$$= \sum_{\mathbf{k}'} \frac{|V_{\mathbf{k}\mathbf{k}'}|^2}{\epsilon_{\mathbf{k}} - \epsilon_{\mathbf{k}'}}.$$

The matrix elements $V_{\mathbf{k}\mathbf{k}'}$ are equal to zero unless $\mathbf{k}'$ is related to $\mathbf{k}$ according to

$$\mathbf{k} - \mathbf{k}' = 2\pi\mathbf{G}.$$

Such a nonvanishing matrix element is equal to $V_{\mathbf{G}}$ in the expansion of V. We then have

$$\epsilon_2 = \frac{2m}{\hbar^2}\sum_{\mathbf{G}} \frac{|V_G|^2}{k^2 - |\mathbf{k}+2\pi\mathbf{G}|^2} = -\frac{m}{2\pi\hbar^2}\sum_{\mathbf{G}} \frac{|V_G|^2}{\mathbf{G}\cdot(\mathbf{k}+\pi\mathbf{G})}. \tag{3.22}$$

The perturbation energy ϵ_2 of a wave increases rapidly as its wave-vector $\mathbf{k}$ approaches the condition

$$\mathbf{G}\cdot\mathbf{k} = -\pi\mathbf{G}^2. \tag{3.23}$$

This is just the Bragg condition for wave reflection given by (1.24). When the condition is satisfied, two states with, respectively, $\mathbf{k}$

and $\mathbf{k}' = \mathbf{k} - 2\pi\mathbf{G}$ are degenerate in the zeroth-order approximation. Standard perturbation treatment shows that ϵ_2 has two values:

$$\epsilon_2 = \pm|V_{\mathbf{G}}|. \tag{3.24}$$

In other words, the energy band $\epsilon(\mathbf{k})$ shows an energy gap at each zone boundary in $\mathbf{k}$ space.

The energy ϵ in this approximation increases by and large in proportion to k^2. In the scheme of reduced wavevectors, in which wavevectors of each zone are shifted into the first zone, considerations can be limited to the first zone, which shows energy bands $\epsilon(\mathbf{k})$ at successively higher energies with an energy gap between adjacent energy bands.

a. Electron Dynamics

Velocity. Substituting the Bloch function into the Hartree–Fock equation, differentiating the equation with respect to the wavevector component k_1, multiplying the result by $\varphi_{\mathbf{k}}^*$, and integrating over space, we get

$$\frac{\partial \epsilon}{\partial k_1} + \frac{i\hbar^2}{m}\int \varphi_{\mathbf{k}}^* \frac{\partial \varphi_{\mathbf{k}}}{\partial x} d\tau + \int \varphi_{\mathbf{k}}^*(-H^{\mathrm{F}} + \epsilon_{\mathbf{k}})\left[\exp(i\mathbf{k}\cdot\mathbf{r})\frac{\partial u_{\mathbf{k}}}{\partial k_1}\right] d\tau = 0. \tag{3.25}$$

Applying Green's theorem, we write the last term on the left-hand side in the following form:

$$\int \exp(i\mathbf{k}\cdot\mathbf{r})\frac{\partial u_{\mathbf{k}}}{\partial k_1}(-H^{\mathrm{F}} + \epsilon_{\mathbf{k}})\varphi_{\mathbf{k}}^* \, d\tau$$
$$- \frac{\hbar^2}{2m}\int \left\{\varphi_{\mathbf{k}}^* \nabla \left[\exp(i\mathbf{k}\cdot\mathbf{r})\frac{\partial u_{\mathbf{k}}}{\partial k_1}\right] - \left[\exp(i\mathbf{k}\cdot\mathbf{r})\frac{\partial u_{\mathbf{k}}}{\partial k_1}\right]\nabla\varphi_{\mathbf{k}}^*\right\} dS.$$

The volume integral is obviously equal to zero. The surface integral vanishes for a surface enclosing an integral number of lattice cells, since the integrand is periodic in the lattice. Therefore,

$$\frac{1}{\hbar}\frac{\partial \epsilon}{\partial k_1} = \frac{\hbar}{mi}\int \varphi_{\mathbf{k}}^* \frac{\partial \varphi_{\mathbf{k}}}{\partial x} d\tau = \frac{\hbar}{2mi}\int \left(\varphi_{\mathbf{k}}^* \frac{\partial \varphi_{\mathbf{k}}}{\partial x} - \varphi_{\mathbf{k}}\frac{\partial \varphi_{\mathbf{k}}^*}{\partial x}\right) d\tau. \tag{3.26}$$

On the right-hand side, we have the familiar expression of the x component of the group velocity $\mathbf{v} = d\langle\mathbf{r}\rangle/dt$, the time-dependent expectation value $\langle \mathbf{r} \rangle$ being given by the time-dependent function. The velocity is then given by

$$\mathbf{v} = \frac{1}{\hbar} \nabla_{\mathbf{k}} \epsilon. \tag{3.27}$$

Acceleration. Consider the effect of an applied constant electric field **E**. The wave equation for the time-dependent wavefunction φ is

$$i\hbar \frac{\partial \varphi}{\partial t} = \left[-\frac{\hbar^2}{2m} \nabla^2 + V - e\mathbf{E} \cdot \mathbf{r} \right] \varphi. \tag{3.28}$$

The solutions can be expressed in the following form:

$$\varphi(r,t) = \sum_m \int a_m(\mathbf{k},t) \varphi_{m\mathbf{k}}(\mathbf{r})\, d\mathbf{k}, \tag{3.29}$$

where $\varphi_{m\mathbf{k}}$ is an eigenfunction in the absence of an applied field, **k** is the reduced wavevector, and m is the index of energy bands. Substitute this expression for φ into Eq. (3.28). Taking into account the nature of the Bloch function, we get by straight analysis

$$\left(\hbar \frac{\partial}{\partial t} + e\mathbf{E} \cdot \nabla_{\mathbf{k}} \right) \sum_m |a_m|^2 = 0.$$

Consider an electron that is initially in some energy band. For sufficiently short times, only this one band is of importance, and we may drop the band index:

$$\hbar \frac{\partial}{\partial t} |a|^2 = -e\mathbf{E} \cdot \nabla_{\mathbf{k}} |a|^2. \tag{3.30}$$

An applied uniform magnetic field **B** introduces two terms into the Hamiltonian; one is linear in **B** and the second is proportional to B^2. The second term can usually be neglected. It can be shown that the magnetic field gives

$$\hbar \frac{\partial}{\partial t} |a|^2 = -\frac{e}{2c} [(\mathbf{v} + \langle \mathbf{v} \rangle) \times \mathbf{B}] \cdot \nabla_{\mathbf{k}} |a|^2$$

where

$$\langle \mathbf{v} \rangle = \int \mathbf{r} \frac{\partial}{\partial t} (\varphi^* \varphi)\, dr. \tag{3.31}$$

Combining the effects of an electric field and a magnetic field, we get

$$\hbar \frac{\partial}{\partial t} |a|^2 = -e \left\{ \mathbf{E} + \frac{1}{2c} [(\mathbf{v} + \langle \mathbf{v} \rangle) \times \mathbf{B}] \right\} \cdot \nabla_{\mathbf{k}} |a|^2. \tag{3.32}$$

It follows that $|a|^2$ is a function of

$$\left\{\mathbf{k} - e\left[\mathbf{E} + \frac{1}{2c}(\mathbf{v} + \langle\mathbf{v}\rangle) \times \mathbf{B}\right] t/\hbar\right\},$$

showing that the applied fields make $\mathbf{k}$ change at a rate of

$$\frac{d\mathbf{k}}{dt} = \frac{e}{\hbar}\left\{\mathbf{E} + \frac{1}{2c}[(\mathbf{v} + \langle\mathbf{v}\rangle) \times \mathbf{B}]\right\}. \tag{3.33}$$

If $|a|^2$ is large only for $\mathbf{k}$ close to a particular wavevector $\mathbf{k}_0$, then $\mathbf{v}$ and $\langle\mathbf{v}\rangle$ are both approximately $\mathbf{v}_{\mathbf{k}_0}$ for sufficiently short times. We have then

$$\frac{d\mathbf{k}}{dt} = \frac{e}{\hbar}\left[\mathbf{E} + \frac{1}{c}\mathbf{v} \times \mathbf{B}\right]. \tag{3.34}$$

Assume the applied fields $\mathbf{E}$ and $\mathbf{B}$ to be sufficiently weak that it is still possible to consider them in terms of the Bloch states. The acceleration $d\langle\mathbf{v}\rangle/dt$ of an electron, a Bloch wave packet, is given by

$$\begin{aligned}\frac{d\mathbf{v}}{dt} &= \frac{d}{dt}\left(\frac{1}{\hbar}\nabla_{\mathbf{k}}\epsilon\right) = \frac{1}{\hbar}\nabla_{\mathbf{k}}\left(\frac{d\epsilon}{dt}\right) = \frac{1}{\hbar}\nabla_{\mathbf{k}}\left(\nabla_{\mathbf{k}}\epsilon \cdot \frac{d\mathbf{k}}{dt}\right)\\ &= \frac{1}{\hbar^2}(\nabla_{\mathbf{k}}\nabla_{\mathbf{k}}\epsilon) \cdot e\left(\mathbf{E} + \frac{1}{c}\mathbf{v} \times \mathbf{B}\right).\end{aligned} \tag{3.35}$$

Since $e[\mathbf{E} + (1/c)\mathbf{v} \times \mathbf{B}]$ is the force experienced by the electron, the tensor

$$m^* = \left[\frac{1}{\hbar^2}\nabla_{\mathbf{k}}\nabla_{\mathbf{k}}\epsilon\right]^{-1} \tag{3.36}$$

is called the effective mass of the electron.

The Concept of Holes. A state not occupied by an electron is considered to be occupied by a hole. The overall behavior of electrons in an energy band is more conveniently considered in terms of holes when there are a limited number of holes in the band. A hole represents a missing electron. The motion of a hole being that of the missing electron, the acceleration of a hole under the force $\mathbf{F}_e$ that would have been experienced by the missing electron is

$$\frac{d\mathbf{v}}{dt} = \frac{1}{\hbar^2}[\nabla_{\mathbf{k}}\nabla_{\mathbf{k}}\epsilon] \cdot \mathbf{F}_e = \frac{1}{\hbar^2}[\nabla_{\mathbf{k}}\nabla_{\mathbf{k}}(-\epsilon)] \cdot (-\mathbf{F}) \equiv \frac{1}{m_{\mathrm{h}}}(-\mathbf{F}_e). \tag{3.37}$$

The effective reciprocal mass m_{h}^{-1} is usually positive, since the missing electrons are likely to be from states close to the top of the energy band. The minus sign in $-\mathbf{F}_e$ shows that the charge to be assigned to a hole is opposite in sign to the electronic charge. An insulator or semiconductor is often said to be n-type (n for negative) or p-type (p for positive), depending upon whether conduction electrons or holes are dominant in its electronic properties.

Oscillator Strength and the f-Sum Rule. Inserting the Bloch function for φ into the expression for $\mathbf{v}$, we get

$$\frac{m}{\hbar^2}\nabla_{\mathbf{k}}\epsilon = \frac{m}{\hbar}\langle\mathbf{v}\rangle = \mathbf{k} + \frac{1}{i}\int u_{\mathbf{k}}^*\nabla u_{\mathbf{k}}\,d\tau.$$

Differentiation of this expression gives

$$\frac{m}{\hbar^2}\frac{\partial^2\epsilon}{\partial k_i^2} = 1 + i\int\left(\frac{\partial u^*}{\partial r_i}\frac{\partial u}{\partial k_i} - \frac{\partial u}{\partial r_i}\frac{\partial u^*}{\partial k_i}\right)d\tau,$$

$$\frac{m}{\hbar^2}\frac{\partial^2\epsilon}{\partial k_i\partial k_j} = i\int\left(\frac{\partial u^*}{\partial r_i}\frac{\partial u}{\partial k_j} - \frac{\partial u}{\partial r_i}\frac{\partial u^*}{\partial k_j}\right)d\tau$$
$$= i\int\left(\frac{\partial u^*}{\partial r_j}\frac{\partial u}{\partial k_i} - \frac{\partial u}{\partial r_j}\frac{\partial u^*}{\partial k_i}\right)d\tau,$$

where i and j are indices of components.

The second derivatives of ϵ can be expressed in terms of the matrix components of the momentum operator $\mathbf{p}$, which connect Bloch states of different energy-band m's for the same $\mathbf{k}$:

$$(m'|p_i|m'') = \frac{\hbar}{i}\int\varphi_{m''}^*\frac{\partial}{\partial r_i}\varphi_{m'}\,d\tau = \frac{\hbar}{i}\int u_{m''}^*\frac{\partial}{\partial r_i}u_{m'}\,d\tau. \tag{3.38}$$

The integral in terms of u is obtained from the integral in terms of the Bloch function. On the other hand, substituting the Bloch function into the Hartree–Fock equation and making appropriate mathematical manipulations, we get

$$\frac{\hbar}{i}\int\varphi_{m''}^*\frac{\partial}{\partial r_i}\varphi_{m'}^*\,d\tau = -\frac{m}{\hbar}(\epsilon_{m''} - \epsilon_{m'})\int u_{m''}^*\frac{\partial}{\partial k_i}u_{m'}\,d\tau.$$

Hence we may write

$$\sum_{m''}{}'\frac{2}{m}\frac{|(m''|p_i|m')|^2}{\epsilon_{m''} - \epsilon_{m'}} = \frac{1}{i}\sum_{m''}\left[\left(\int u_{m''}^*\frac{\partial u_{m'}}{\partial k_i}d\tau\right)\left(\int u_{m''}\frac{\partial u_{m'}^*}{\partial r_i}d\tau\right)\right.$$
$$\left. - \left(\int u_{m''}\frac{\partial u_{m'}^*}{\partial k_i}d\tau\right)\left(\int u_{m''}^*\frac{\partial u_{m'}}{\partial r_i}d\tau\right)\right].$$

A term with $m'' = m'$ has been added to the summation on the right-hand side. This term is equal to zero since

$$\frac{\partial}{\partial k_i} \int |u_{m'}|^2 \, d\tau = 0.$$

Now the u_m's of various energy bands and the same $\mathbf{k}$ are a complete set of orthogonal functions, and therefore can be used to expand an arbitrary function. It follows that

$$\sum_{m''} \left(\int u_{m''}^* \frac{\partial u_{m'}}{\partial k_i} d\tau \right) \left(\int u_{m''} \frac{\partial u_{m'}^*}{\partial r_i} d\tau \right)$$
$$= \int \frac{\partial u_{m'}}{\partial k_i} \left(\sum_{m''} u_{m''}^* \int u_{m''} \frac{\partial u_{m'}^*}{\partial r_i} d\tau \right) d\tau = \int \frac{\partial u_{m'}}{\partial k_i} \frac{\partial u_{m'}^*}{\partial r_i} d\tau,$$

since the bracketed quantity in the integrand of the second expression is $\partial u_{m'}^* / \partial r_i$ expanded in terms of u_m. Consequently,

$$\sum_{m''}{}' (f_{m''m'})_{ii} \equiv \sum_{m''}{}' \frac{2}{m} \frac{|(m''|p_i|m')|^2}{\epsilon_{m''} - \epsilon_{m'}}$$
$$= i \int \left(\frac{\partial u_{m'}}{\partial r_i} \frac{\partial u_{m'}^*}{\partial k_i} - \frac{\partial u_{m'}^*}{\partial r_i} \frac{\partial u_{m'}}{\partial k_i} \right) d\tau. \tag{3.39}$$

Analogous analysis shows

$$\sum_{m''}{}' (f_{m''m'})_{ij}$$
$$\equiv \sum_{m''}{}' \frac{1}{m} \frac{(m'|p_i|m'')(m''|p_j|m') + (m'|p_j|m'')(m''|p_i|m')}{\epsilon_{m''} - \epsilon_{m'}}$$
$$= i \left(\frac{\partial u_{m'}}{\partial r_j} \frac{\partial u_{m'}^*}{\partial k_i} - \frac{\partial u_{m'}^*}{\partial r_j} \frac{\partial u_{m'}}{\partial k_i} \right) d\tau. \tag{3.40}$$

The tensor quantity $f_{m''m'}$ defined above is the oscillator strength connecting energy band m' with band m'' for states of a specified wavevector $\mathbf{k}$.

The relations between components of the m/m^* tensor and the oscillator strengths $f_{m''m'}$ are

$$\frac{m}{\hbar^2} \frac{\partial^2 \epsilon_{m'}}{\partial k_i \partial k_j} = -\sum_{m''}{}' (f_{m''m'})_{ij}, \quad \frac{m}{\hbar^2} \frac{\partial^2 \epsilon_{m'}}{\partial k_i^2} = 1 - \sum_{m''}{}' (f_{m''m'})_{ii}. \tag{3.41}$$

The latter is sometimes called the f-sum rule. $(m/\hbar^2)(\partial^2 \epsilon_{m'} / \partial k_i^2)$ plays the role of $(f_{m'm'})_{ii}$.

3. GROUP REPRESENTATION

The representation of a group G is a group D of square matrices that is homomophric with G, the combination of matrices being matrix multiplication. To each element A of G corresponds a matrix $D(A)$; to a combination of elements $AB = C$ corresponds the matrix multiplication $D(A)D(B) = D(C)$. The number of columns or rows of a matrix is called the dimension of the representation.

The identity element E of a group is represented by a unit matrix: $(E) = \mathbf{1}$, since $EA = A$ for any element A of the group. Furthermore, we have $AA^{-1} = E$ for any element A and the inverse element A^{-1}. Therefore,

$$D(A)D(A^{-1}) = \mathbf{1}, \qquad \text{or} \qquad D(A^{-1}) = \{D(A)\}^{-1}.$$

For the determinants of the matrices, we get

$$|D(A^{-1})| = |D(A)|^{-1} = \frac{1}{|D(A)|}. \tag{3.42}$$

This result shows that a representation must consist of nonsingular matrices, each having a nonvanishing determinant.

Equivalent Representations. The similarity transformation of a matrix M by means of a nonsingular square matrix P of the same dimensions is $P^{-1}MP$. Subjecting every matrix of a representation to the same similarity transformation, we get another representation, since the multiplication relations of the matrices are not altered by such a transformation. The two representations are considered to be equivalent. Evidently, the matrices of equivalent representations have a common dimension. Any representation consisting of nonsingular matrices can be transformed by a similarity transformation into a representation with unitary matrices. Two equivalent unitary representations can be transformed into each other by using a unitary matrix.

Reducible Representations. A representation D of a group can be formed by combining two representations D_1 and D_2: Join the two matrices for each element along the diagonal, and fill out each combined matrix with zeros. The representation D is said to be reducible and is written as $D = D_1 + D_2$. A reducible representation may not be given in a form easily recognizable as a combination of two or more representations, but it must be possible to bring it to the proper form by an appropriate similarity transformation.

Irreducible Representation. An irreducible representation is one that cannot be reduced. It can be shown that the number of irreducible representations of a group is equal to the number of classes of group elements. Let n_λ be the dimension of the irreducible representation λ, and let h be the number of elements in the group. The following equation holds:

$$\sum_{\lambda} n_\lambda^2 = h. \tag{3.43}$$

Schur's lemmas. Let the matrix P of the order $n_2 \times n_1$ relate two irreducible representations, D_1 of dimension n_1 and D_2 of dimension n_2, according to

$$PD_1(T) = D_2(T)P \qquad \text{or} \qquad PD_1(T)P^{-1} = D_2(T)$$

for every element of the group. The following lemmas are useful and easily understandable.

(i) $P = 0$ if D_1 and D_2 are not equivalent.
(ii) $P = 0$ or $|P| \neq 0$ if D_1 and D_2 are equivalent.
(iii) $P = 0$ or $P = a\mathbf{1}$ if D_1 and D_2 are the same representation, where a is a constant and $\mathbf{1}$ is the unit matrix. In other words, a matrix that commutes with every matrix of an irreducible representation is a multiple of the unit matrix. The converse is also true: If the only matrix that commutes with every matrix of a representation is a multiple of the unit matrix, then the representation is irreducible.

Characters. The sum χ of the diagonal elements of $D(T)$ is called the character of the group element T in the representation D:

$$\chi(T) = \sum_i D_{ii}(T). \tag{3.44}$$

Since the identity element is always represented by a unit matrix, the character of the identity element is always equal to the dimension of representation.

Mathematically,

$$\begin{aligned}\sum_i [SD(T)S^{-1}]_{ii} &= \sum_{ijk} S_{ij}D_{jk}(T)S_{ki}^{-1} = \sum_{ijk} S_{ki}^{-1}S_{ij}D_{jk}(T) \\ &= \sum_{jk} \delta_{kj}D_{jk}(T) = \chi(T),\end{aligned} \tag{3.45}$$

where S is any square matrix of the same order as that of a matrix in the representation D. Two conclusions follow from this relation:

1. The characters of all elements in one class are the same, since S may be the matrix representing any element of the group, making the left-hand side of the relation the character of any element conjugate to T.
2. The characters of equivalent representations are the same, since S may be the matrix connecting two equivalent representations by a similarity transformation.

Thus each representation has specified characters of the elements. The character table of a group lists all the nonequivalent representations, each with its characters for the various classes of elements.

Orthogonality of Irreducible Representations. It can be shown that two irreducible unitary representations λ and μ have the orthogonality relation

$$\sum_T D^{\lambda *}_{\alpha\beta}(T) D^{\mu}_{ij}(T) = \frac{h}{n_\lambda}\delta_{\alpha i}\delta_{\beta j}\delta_{\lambda\mu}. \tag{3.46}$$

The connotation of orthogonality may be made more apparent by considering the case of $\mu = \lambda$, for which

$$\sum_T D^*_{\alpha\beta}(T) D_{ij}(T) = \frac{h}{n}\delta_{\alpha i}\delta_{\beta i}.$$

For each pair $\alpha\beta$, there are h factors: $D_{\alpha\beta}(T_1), D_{\alpha\beta}(T_2), \ldots, D_{\alpha\beta}(T_h)$. Consider these factors the components of a vector D in h-dimensional space. There are n^2 such vectors. The above equation expresses the orthogonality of these vectors, each of which has a Hermitian length of $(h/n)^{-1/2}$.

Orthogonality Relation for the Characters. Put $\alpha = \beta$ and $i = j$ in the orthogonality relation (3.46) for two irreducible unitary representations. Summing over α, we get

$$\sum_T \chi^{\lambda^*}(T)\chi^{\mu}(T) = h\delta_{\lambda\mu}. \tag{3.47}$$

Let s be the index of class and h_s be the number of elements in the class. The above relation can be written as

$$\sum_s h_s \chi_s^{\lambda^*} \chi_s^{\mu} = h\delta_{\lambda\mu}. \tag{3.48}$$

Since the characters are not affected by similarity transformations of representation, the above relation, known as the orthogonality relation, holds irrespective of whether the representation is unitary.

For the same irreducible representation,

$$\sum_s h_s |\chi_s|^2 = h, \tag{3.49}$$

which may be used to test the irreducibility of the representation.

Define a matrix U with components

$$U_{s\lambda} = (h_s/h)^{1/2} \chi_s^\lambda.$$

The matrix is square, since the number of representations is equal to the number of classes. The orthogonality relation can be written as

$$\sum_s U_{\lambda s}^+ U_{s\mu} = \delta_{\lambda\mu},$$

showing that U is a unitary matrix:

$$U^+ U = E = UU^+.$$

It follows that relation (3.48) leads to

$$\sum_\mu (h_s/h)^{1/2} \chi_s^\mu (h_{s'}/h)^{1/2} \chi_{s'}^{\mu^*} = \delta_{ss'}, \tag{3.50}$$

which is called the second orthogonality relation for characters.

Decomposition of a Reducible Representation. In a representation, the character $\chi(T)$ is given by

$$\chi(T) = \sum_\lambda c_\lambda \chi^\lambda(T), \tag{3.51}$$

where c_λ is the number of times the irreducible representation D^λ is contained in the representation. Multiply this equation by the character $\chi^{\mu^*}(T)$ in an irreducible representation D^μ, repeat the operation for every element T, and sum over T. In view of orthogonality condition (3.47) or (3.48), we get

$$c_\mu = \frac{1}{h} \sum_T \chi^{\mu^*}(T)\chi(T) = \frac{1}{h} \sum_s h_s \chi_s^{\mu^*} \chi_s. \tag{3.52}$$

This relation may be used to determine the number of times a known irreducible representation is contained in a given representation.

Direct-Product Representation. We have from matrix algebra that the direct product of corresponding matrices of two different representations constitutes a representation of the group:

$$D^{\lambda\times\mu}(R) = D^{\lambda}(R)D^{\mu}(R). \tag{3.53}$$

With D^{λ} having dimension m and D^{μ} having dimension n, the dimension of the direct-product representation $D^{\lambda\times\mu}$ is mn. Since

$$\sum_{i,j} D_{ij,ij} = \sum_{i,j} D_{ii}D_{jj} = \left(\sum_{i} D_{ii}\right)\left(\sum_{j} D_{jj}\right),$$

we have

$$\chi^{\lambda\times\mu}(R) = \chi^{\lambda}(R)\chi^{\mu}(R). \tag{3.54}$$

A direct-product representation is in general reducible. According to (3.51), the number of times that an irreducible representation ν is contained in the direct-product representation is

$$c_{\nu}^{\lambda\times\chi} = h^{-1}\sum_{R}\chi^{\lambda\times\mu}(R)\chi^{\nu*}(R) = h^{-1}\sum_{s} h_s\chi_s^{\lambda}\chi_s^{\mu}\chi_s^{\nu*}. \tag{3.55}$$

Direct-Product Group. Consider two different groups,

$$G_\alpha = E, A_2, A_3, \ldots, A_{h_1} \qquad \text{and} \qquad G_\beta = E, B_2, B_3, \ldots, B_{h_2}.$$

Every element of one group commutes with all elements of the other. The direct-product group is

$$G_\alpha \times G_\beta = E, A_2, \ldots, A_{h_1}, B_2, A_2B_2, \ldots, A_{h_1}B_2, \ldots, A_{h_1}B_{h_2}, \tag{3.56}$$

which has h_1h_2 elements. The product $G_1 \times G_2$ is indeed a group, since

$$(A_iB_j)(A_{i'}B_{j'}) = (A_iA_{i'})(B_jB_{j'}),$$

showing that a combination of two elements in $G_1 \times G_2$ is a combination of one element of G_1 with an element of G_2 and is therefore an element of $G_1 \times G_2$. It also follows from the commutativity of the A's with the B's that the number of classes of the product group is simply the product of the numbers of classes of the component groups.

As to the matrices and characters of the representations, we have, from matrix algebra,

$$D^{(\alpha\times\beta)}(A_iB_j) = D^{\alpha}(A_i)D^{\beta}(B_j)$$

and

$$\chi^{(\alpha\times\beta)}(A_iB_j) = \chi^{\alpha}(A_i)\chi^{\beta}(B_j). \tag{3.57}$$

It can be shown by using Schur's lemmas that the direct product of two irreducible representations of the component groups forms an irreducible representation of the direct-product group, and that the above equation of characters can be used to obtain the characters of all the irreducible representations of the product group.

The Regular Representation D^{reg}. Consider the multiplication table of a given group. The top row lists the elements $E, A, B, \ldots$. The inverses of the elements $E^{-1} = E, A^{-1}, B^{-1}, \ldots$ constitute the same group, and they will be listed in that order in the left column of the table. The diagonal elements in this table are obviously the identity element E. The matrix $D^{\text{reg}}(F)$ representing the element F is obtained from this multiplication table by replacing the F's with unity and putting a zero everywhere else. For example,

$$D^{\text{reg}}_{MK^{-1}}(F) = \begin{cases} 1, & \text{if } MK^{-1} = F \\ 0, & \text{if } MK^{-1} \neq F. \end{cases} \tag{3.58}$$

We show that D^{reg} is a representation of the group by proving that

$$D^{\text{reg}}(BC) = D^{\text{reg}}(B)D^{\text{reg}}(C)$$

or

$$D^{\text{reg}}_{MK^{-1}} = \sum_{J^{-1}} D^{\text{reg}}_{MJ^{-1}}(B)D^{\text{reg}}_{JK^{-1}}(C), \tag{3.59}$$

where B, C, M, K, and J are any elements of the group. In fact, a nonvanishing term equal to unity occurs in the sum on the right-hand side only when

$$MJ^{-1} = B \qquad \text{and} \qquad JK^{-1} = C,$$

which amounts to

$$BC = MJ^{-1}JK^{-1} = MK^{-1}.$$

The last condition gives also

$$D^{\text{reg}}_{MK^{-1}}(BC) = 1.$$

Thus D^{reg} is proved to be a representation of the group.

From the definition of D^{reg}, we have

$$\chi^{\text{reg}}(E) = \text{dimension of } D^{\text{reg}} = h, \text{ order of the group,}$$

and

$$\chi^{\text{reg}}(F) = 0 \qquad \text{for} \quad F \neq E. \tag{3.60}$$

According to (3.51), any irreducible representation D^μ is contained in the D^{reg} representation c_μ^{reg} times, which is given by

$$c_\mu^{\text{reg}} = \frac{1}{h}\sum_F \chi^{\mu *}(F)\chi^{\text{reg}}(F) = \frac{1}{h}\chi^{\mu *}(E)h = n_\mu. \tag{3.61}$$

The order of the group is

$$h = \sum_\mu c_\mu^{\text{reg}} n_\mu = \sum_\mu n_\mu^2 = \sum_\mu |\chi^\mu(E)|^2. \tag{3.62}$$

This relation can be used to test whether a given set of irreducible representations is a complete set.

a. Determination of Irreducible Representations

The matrices of the regular representation can be easily obtained. The matrices of all the irreducible representations can be found if the matrices of D^{reg} can be reduced. There is no general method for accomplishing this task. There is, however, a general method for determining the characters of the irreducible representations, which is outlined in the following.

Let C_i denote the collection of elements in the class indicated by the subscript i. We know that the whole collection C_s is contained in the collection C_iC_j if any one element of C_s is contained in C_iC_j. Therefore,

$$C_iC_j = \sum_s c_{ij,s}C_s. \tag{3.63}$$

The following relation can be proved for an irreducible representation of the group:

$$h_i\chi_i h_j\chi_j = \chi(E)\sum_s c_{ij,s}h_s\chi_s. \tag{3.64}$$

Proof: Let X be any element of the group. We have

$$X^{-1}C_iX = C_i \qquad \text{or} \qquad C_iX = XC_i.$$

The matrix $D(X)$ represents the element X in the irreducible representation. Let $\mathcal{D}_i$ be the sum of matrices that represent the elements in class i. We have

$$\mathcal{D}_i D(X) = D(X)\mathcal{D}_i,$$

since matrix multiplication is linear and matrix addition is commutative. According to Schur's lemma (iii),

$$\mathcal{D}_i = a_i E, \tag{3.65}$$

where a_i is a constant and E is the unit matrix. The right-hand side of (3.63) written in matrix notation is

$$\sum_s c_{ij,s} \mathcal{D}_s = \sum_s c_{ij,s} a_s E = E \sum_s c_{ij,s} a_s.$$

The left-hand side of (3.63),

$$C_i C_j = X^{-1} C_i X X^{-1} C_j X,$$

gives the matrix expression

$$\begin{aligned} &D(X^{-1})\mathcal{D}_i D(X) D(X^{-1}) \mathcal{D}_j D(X) \\ &\quad = D(X^{-1})\mathcal{D}_i \mathcal{D}_j D(X) = a_i a_j D(X^{-1}) E E D(X) \\ &\quad = a_i a_j D(X^{-1}) D(X) = a_i a_j E. \end{aligned}$$

Therefore (3.63) leads to

$$a_i a_j = \sum_s c_{ij,s} a_s. \tag{3.66}$$

It follows from (3.65) that

$$\mathrm{Tr}(\mathcal{D}_i) = a_i \chi(E)$$

where $\chi(E)$ is the dimension of the representation D. On the other hand, $\mathcal{D}_i$ as defined has

$$\mathrm{Tr}(\mathcal{D}_i) = h_i \chi_i.$$

The above two equations give

$$a_i = h_i \chi_i / \chi(E). \tag{3.67}$$

Inserting the last equation into (3.66), we get Eq. (3.64).

With the notation $\chi_1 = \chi(E)$, (3.64) may be written as

$$\frac{h_i \chi_i}{\chi_1} \frac{h_j \chi_j}{\chi_1} = \sum_s c_{ij,s} \frac{h_s \chi_s}{\chi_1}. \tag{3.68}$$

A number of equations like this provide the determination of the $(h_i\chi_i/\chi_1)$'s or (χ_i/χ_1)'s. The value of χ_1 can be determined from the orthogonality relation (3.49) of the characters:

$$\chi_1^2 \sum_s \frac{1}{h_s} \left| \frac{h_s \chi_s}{\chi_1} \right|^2 = h. \tag{3.69}$$

In the following, we outline briefly the formal solution of (3.68). Multiplying each side of (3.68) by an arbitrary coefficient A_i and summing over i, we get

$$\xi \frac{h_j \chi_j}{\chi_1} = \sum_{i,s} A_i c_{ij,s} \frac{h_s \chi_s}{\chi_1}, \tag{3.70}$$

where

$$\xi = \sum_{i=1}^{r} \frac{A_i h_i \chi_i}{\chi_1}. \tag{3.71}$$

There are r homogeneous equations for the $r-1$ unknowns $h_i\chi_i/\chi_1$. A solution requires

$$\begin{vmatrix} \sum_i A_i c_{i1,1} - \xi & \sum_i A_i c_{i1,2} & \cdots & \sum_i A_i c_{i1,r} \\ \sum_i A_i c_{ir,1} & \sum_i A_i c_{ir,2} & \cdots & \sum_i A_i c_{ir,r} - \xi \end{vmatrix} = 0. \tag{3.72}$$

The determinant is a homogeneous equation of the rth degree in ξ, with rational integral coefficients. It should be a product of r linear factors containing ξ. Each linear factor

$$\xi - \sum_i \alpha_i A_i = A_i \frac{h_i \chi_i}{\chi_1} - \sum_i \alpha_i A_i \tag{3.73}$$

gives a solution of ξ. Since the A_i's are arbitrary, each solution of ξ gives a set of (χ_i/χ_1)'s:

$$h_i \chi_i / \chi_1 = \alpha_i. \tag{3.74}$$

One method of finding a linear factor of the determinant is to find a linear combination of such rows, which consists of a row of elements identical in absolute magnitude, since any row of a determinant may be replaced by a combination of some rows. The common factor of the elements in the combination row may be taken out of the determinant, and hence it is a factor of the product.

b. Symmetry and Quantum Mechanics

Consider a linear transformation of spatial coordinates:

$$x_i = T_{ij}x'_j.$$

An operator or a function can be expressed in either system of coordinates x_i or x'_i. Consider the Hamiltonian operator:

$$H(x_i) = H(T_{ij}x'_j) = H'(x'_i).$$

If

$$H'(x'_i) = H(x'_i), \qquad \text{and consequently} \qquad H(x_i) = H(x'_i), \quad (3.75)$$

then T is a symmetry transformation of the Hamiltonian. Applying T to a function $H(x_i)F(x_i)$, we get

$$TH(x_i)F(x_i) = H(x_i)TF(x_i). \quad (3.76)$$

It may be said in general that a symmetry transformation of an operator commutes with the operator.

A Schrödinger equation operated on by a symmetry operation yields

$$TH(x_i)\varphi(x_i) = T\epsilon\varphi(x_i) = \epsilon T\varphi(x_i), \quad (3.77)$$

showing that the two eigenfunctions $\varphi(x_i)$ and $T\varphi(x_i)$ are degenerate in energy. Using a subscript to indicate a degenerate eigenfunction, we have

$$T\varphi_m(x) = \sum_n D_{nm}(T)\varphi_n(x), \quad (3.78)$$

where the summation is over the set of degenerate eigenfunctions. The complete set of symmetry transformations is a group, and $D(T)$ is the matrix representing the element T. The functions φ form a basis for the representation D; they are said to be the base vectors of the vector space for linear combinations of the degenerate functions. The dimension of the space is equal to the number of independent degenerate functions, and it is equal to the dimension of representation. The vector space or the group representation for an energy level is irreducible apart from the so-called accidental degeneracy; symmetry considerations require that different vector spaces belong to different energy levels. Finally, it should be borne in mind that the above statements, particularly those about irreducibility, depend upon completeness of the symmetry transformations being considered.

4. SYMMETRY AND ELECTRONIC ENERGY BANDS

a. Disregarding the Effect of Electron Spin

The space group of a crystalline solid is the symmetry group of the one-electron Hamiltonian for transformations of spatial coordinates and their derivatives. Primitive translations constitute a subgroup that is Abelian and therefore has one-dimensional irreducible representations. In the one-electron approximation, each Bloch function with its reduced wavevector **k** is the basis function of a irreducible representation of the translational subgroup.

A space group of a lattice is a direct-product group of translations and the rest of the elements. Bloch functions of various **k**'s may be taken as basis functions for a representation of the space group. For a given **k**, there may exist several degenerate eigenfunctions. Since a translation does not change **k** and does not change one degenerate eigenfunction to another, each lattice translation is represented by a diagonal matrix; the number of diagonal elements associated with one **k** is equal to the number of degenerate eigenfunctions of the **k**.

An element of the space group that is not purely a lattice translation changes the **k** of Bloch function. A Bloch function $\varphi_{\mathbf{k}}(\mathbf{r})$ is characterized by the relation

$$\varphi(\mathbf{r}+\mathbf{a}_n) = \exp(i\mathbf{k}\cdot\mathbf{a}_n)\varphi(r),$$

where $\mathbf{a}_n$ is a lattice vector. A coordinate transformation that is purely a translation by **t** gives

$$\varphi(\mathbf{r}+\mathbf{t}+\mathbf{a}_n) = \exp(i\mathbf{k}\cdot\mathbf{a}_n)\varphi(\mathbf{r}+\mathbf{t}),$$

showing that the wavevector is not changed. On the other hand, a coordinate transformation involving some rotation gives

$$\varphi(f(\mathbf{r}'+\mathbf{a}_n')) = \exp(i\mathbf{k}\cdot\mathbf{a}_n)\varphi(f(\mathbf{r}')),$$

in which $\mathbf{r} = f(\mathbf{r}')$ and $\mathbf{r}+\mathbf{a}_n = f(\mathbf{r}'+\mathbf{a}_n') = f(\mathbf{r}')+f(\mathbf{a}_n')$ characterize the transformation. The equation can be written as

$$\varphi'(\mathbf{r}'+\mathbf{a}_n')\exp(i\mathbf{k}\cdot\mathbf{a}_n)\varphi'(\mathbf{r}')$$

or

$$\varphi'(\mathbf{r}'+\mathbf{a}_n') = \exp(i\mathbf{k}'\cdot\mathbf{a}_n')\varphi'(\mathbf{r}'). \tag{3.79}$$

We see that the wavevector is changed from **k** to **k**$'$, which is derived from **k** by applying the transformation to the **k** space. Therefore, symmetry operations other than pure translations are not represented by diagonal matrices.

***Group of Vector* k.** Consider symmetry operations other than pure translations, which leave the wavevector **k** invariant. A screw axis or a glide plane involves a translation that is a fraction of a primitive translation. Therefore, if the operations for invariant **k** contain screw axes and/or glide planes, then a group formed from these operations will involve some lattice translations as its elements. These translations constitute an invariant subgroup $T^{\mathbf{k}}$ of the group $G^{\mathbf{k}}$ for vector **k**. Each translation being represented by a diagonal matrix, it is sufficient to consider the factor group $G^{\mathbf{k}}/T^{\mathbf{k}}$ for the effect of symmetry; for $\mathbf{k} = 0$, $G_{\mathbf{k}}/T_{\mathbf{k}}$ is just the point group of the crystal. In the absence of screw axes and glide planes, $G^{\mathbf{k}}$ itself is to be considered, and $G^{\mathbf{k}}$ is a subgroup of the point group of the crystal.

Each irreducible representation of $G^{\mathbf{k}}/T^{\mathbf{k}}$ corresponds to a state of the wavevector **k**. The dimension of the representation is equal to the degeneracy of the type of state. There are various energy levels of each type; each energy level is associated with a distinct $u_{\mathbf{k}}(\mathbf{r})$ or a distinct set of degenerate $u_{\mathbf{k}}(\mathbf{r})$'s of a Bloch function.

***Star of* k.** Some of the symmetry operations that include a rotation or a reflection do not leave **k** invariant. The **k**'s generated by such operations constitute the star of **k**. Bloch functions for all the **k**'s are degenerate in energy.

The order of the group $G^{\mathbf{k}}/T^{\mathbf{k}}$ and number of components in the star of **k** depend on **k**. For a **k** more symmetrical in the Brillouin zone, the star is simpler and the order of $G^{\mathbf{k}}/T^{\mathbf{k}}$ is larger.

Compatibility Relations. Consider two neighboring wavevectors $\mathbf{k}'$ and $\mathbf{k}''$. $\mathbf{k}'$ has higher symmetry, and the group of $\mathbf{k}'$ has to contain the group of $\mathbf{k}''$ as a subgroup. Each representation D' can be resolved into D'' representations. Since $\varphi_{\mathbf{k}}(\mathbf{r})$ varies continuously with **k** in an energy band, a particular D'' must be a resolved component of D' and hence is compatible only with certain D''s. By the same token, the representation remains the same as **k** varies if the symmetry of **k** does not change. The compatibility relation of representations for various **k**'s is very effective for tracing the structure of an energy band. Some uncertainty is introduced when two bands cross in energy.

b. Symmetry and Spin

We have considered the Hamiltonian to be independent of the electron spin; therefore the Hamiltonian is invariant under symmetry transformations of the spatial coordinates alone. In fact, the electron spin does enter the Hamiltonian in terms of spin–orbit interaction:

$$H_{1s} = \frac{\hbar}{2m^2c^2}\sum_i S_i \cdot [\nabla V \times p_i]$$
$$- \frac{e^2}{mc^2}\sum_{i\neq j}\sum_j \left[\frac{(\mathbf{r}_i - \mathbf{r}_j)\times(\mathbf{v}_i - \mathbf{v}_j)}{r_{ij}^3} - \frac{1}{2}\frac{(\mathbf{r}_i - \mathbf{r}_j)\times \mathbf{v}_i}{r_{ij}^3}\right]\cdot \mathbf{s} \tag{3.80}$$

and direct magnetic interaction of the electron magnetic moments

$$H_{ss} = \frac{e^2}{m^2c^2}\sum_{i<j}\sum_j \left[\frac{\mathbf{s}_i \cdot \mathbf{s}_j}{r_{ij}^3} - 3\frac{(\mathbf{s}_i \cdot \mathbf{r}_{ij})(\mathbf{s}_j \cdot \mathbf{r}_{ij})}{r_{ij}^5}\right]. \tag{3.81}$$

$\mathbf{s}$ is the operator of spin angular momentum, and the subscripts i and j are electron indices. For a Hamiltonian containing spin, a symmetry operation is a simultaneous transformation of the spatial coordinates and the spin coordinates.

Full Rotation Group. We discuss this group briefly, insofar as it is essential for our problem concerning spin. Let $R(\alpha, \boldsymbol{\xi})$ denote the coordinate transformation corresponding to a rotation through an angle α about an axis $\boldsymbol{\xi}$. The group under consideration consists of transformations for various α's and $\boldsymbol{\xi}$'s, both of which vary continuously, and therefore it is also known as the continuous rotation group. The infinitesimal rotation operator I_{ξ} is defined by

$$\lim_{\alpha\to 0}\frac{R(\alpha,\boldsymbol{\xi}) - 1}{\alpha} = iI_{\xi}. \tag{3.82}$$

In terms of I_{ξ}, a rotation operator $R(\alpha, \boldsymbol{\xi})$ is expressed by

$$R(\alpha,\boldsymbol{\xi}) = \lim_{n\to\infty}(1 + i\tfrac{\alpha}{n}I_{\xi})^n = \exp(i\alpha I_{\xi}).$$

I_{ξ} can be expressed through the infinitesimal rotation operators I_x, I_y, I_z about the three Cartesian coordinate axes:

$$I_{\xi} = lI_x + mI_y + nI_z, \tag{3.83}$$

l, m, n being the direction cosines of $\boldsymbol{\xi}$ in the Cartesian coordinate system. Consequently, any rotation can be expressed in terms of I_x,

I_y, and I_z. Consideration of the geometry of rotation leads to the following relations:

$$iI_x = I_y I_z - I_z I_y, \quad iI_y = I_z I_x - I_x I_z, \quad \text{and} \quad iI_z = I_x I_y - I_y I_x. \tag{3.84}$$

Introducing

$$I_+ = I_x + iI_y \qquad \text{and} \qquad I_- = I_x - iI_y, \tag{3.85}$$

we get three operators I_+, I_-, I_z, which have the following relations:

$$\begin{aligned} I_+ &= I_z I_+ - I_+ I_z, \qquad I_- = I_- I_z - I_z I_-, \qquad \text{and} \\ I_z &= \tfrac{1}{2}(I_+ I_- - I_- I_+). \end{aligned} \tag{3.86}$$

In terms of these operators, the operator I^2 is given by

$$I^2 \equiv I_x^2 + I_y^2 + I_z^2 = \tfrac{1}{2}(I_+ I_- + I_- I_+) + I_z^2. \tag{3.87}$$

Axial Rotation Group. Rotations through various angles about a fixed axis constitute a subgroup of the full rotation group. Let ϕ denote the angle of rotation. The representation of the group gives

$$D(\phi_1)D(\phi_2) = D(\phi_1 + \phi_2).$$

Since the elements commute, the group is Abelian and has one-dimensional representations. The character of a representation m is given by

$$\chi_m(\phi) = e^{im\phi}. \tag{3.88}$$

Let u_m denote a base vector of the mth representation,

$$R(\phi, \mathbf{z})u_m = e^{im\phi}u_m.$$

Then

$$\begin{aligned} I_z u_m &= m u_m, \\ I_z(I_+ u_m) &= (I_+ I_z + I_+)u_m = (m+1)(I_+ u_m), \\ I_z(I_- u_m) &= (I_- I_z - I_-)u_m = (m-1)(I_- u_m). \end{aligned} \tag{3.89}$$

We see that $I_+ u_m$ and $I_- u_m$ belong to the $(m+1)$th and $(m-1)$th representations, respectively.

Let j be the highest value of m; that is,

$$I_+ u_j = 0 \qquad \text{for} \quad u_j \neq 0.$$

It can be shown that $-j-1$ is then the lowest value of m; that is,

$$I_- u_{(-j-1)} = 0 \qquad \text{for} \quad u_{(-j-1)} \neq 0.$$

There are $2j+1$ u_m's, and $2j+1$ is an integer. Therefore,

$$j \text{ must be an integer or half an odd integer.} \tag{3.90}$$

The u_m's of a given j transform according to

$$\begin{aligned} I_+ u_m^{(j)} &= [j(j+1) - m(m+1)]^{1/2} u_{m+1}^{(j)}, \\ I_- u_m^{(j)} &= [j(j+1) - m(m-1)]^{1/2} u_{m-1}^{(j)}, \\ I_z u_m^{(j)} &= m u_m^{(j)}, \\ I^2 u_m^{(j)} &= j(j+1) u_m^{(j)}. \end{aligned} \tag{3.91}$$

Irreducible Representations of the Full Rotation Group. The $u_m^{(j)}$'s of a fixed j transform into each other under the operations of I_+, I_-, and/or I_z. Each element of the full rotation group is composed of the three infinitesimal operators. Therefore the $u_m^{(j)}$'s of a fixed j form base vectors of a representation $D^{(j)}$ of the full rotation group, and the representation is irreducible. A rotation of 2π about a symmetry axis is equivalent to no rotation. According to (3.88), the m's must be integers, and j must be an integer. However, it is shown in the following that j can be half an odd integer due to the nature of the electron.

Electron Spin. A particle may have some inherent degree of freedom, or spin, that gives it direction-dependent properties. In the case of orbital motion, an infinitesimal rotation can be easily seen to correspond to an angular momentum $\mathbf{L}$, for example,

$$I_z = -i\left(x\frac{\partial}{\partial y} - y\frac{\partial}{\partial x}\right) = \frac{L_z}{\hbar}. \tag{3.92}$$

Analogously, the spin is associated with a spin angular momentum $\mathbf{S} = \hbar \mathbf{I}_s$ related to the infinitesimal rotation $\mathbf{I}_s$ of the spin coordinate $\boldsymbol{\sigma}$. For an electron, the spin function is characterized by its value for two different $\boldsymbol{\sigma}$'s. Each $\boldsymbol{\sigma}$ is characterized by its components on two orthogonal axes 1 and 2 in the spin space. Consider two spin functions, u_a and u_b:

$$u_a(\sigma_1 = 1, \sigma_2 = 0) = 1, \qquad u_a(\sigma_1 = 0, \sigma_2 = 1) = 0,$$

and

$$u_b(\sigma_1 = 1, \sigma_2 = 0) = 0, \qquad u_b(\sigma_1 = 0, \sigma_2 = 1) = 1. \tag{3.93}$$

These functions written in matrix form,

$$u_a = \begin{vmatrix} 1 \\ 0 \end{vmatrix} \quad \text{and} \quad u_b = \begin{vmatrix} 0 \\ 1 \end{vmatrix}, \tag{3.94}$$

are called spinors. According to Pauli, these functions or spinors are eigenfunctions of the spin angular momentum along a direction $\mathbf{z}$:

$$S_z u_a = \tfrac{1}{2}\hbar u_a, \qquad S_z u_b = -\tfrac{1}{2}\hbar u_b. \tag{3.95}$$

Any spin function can be expressed in terms of the two orthogonal functions u_a and u_b. Therefore the two functions can serve as base vectors for a representation of the full rotation group. For the spin to possess a directional property, it should not be invariant under all rotations as would be the case for the $D^0 + D^0$ representation; hence the full rotation group must have the $D^{1/2}$ representation. The two values of m for $j = \frac{1}{2}$ are $m = \frac{1}{2}$ and $m = -\frac{1}{2}$. According to (3.91),

$$I_z u_+^{(1/2)} = \tfrac{1}{2} u_+^{(1/2)} \quad \text{and} \quad I_z u_-^{(1/2)} = -\tfrac{1}{2} u_-^{(1/2)}, \tag{3.96}$$

where the subscript + or − is an abbreviation of $+\frac{1}{2}$ or $-\frac{1}{2}$. Comparison of (3.95) and (3.96) shows that the spin functions u_a and u_b correspond to u_+ and u_-, respectively. In the representation $D^{1/2}$,

$$\begin{aligned}
&D^{1/2}(I_z) = \frac{1}{2}\begin{vmatrix} 1 & 0 \\ 0 & -1 \end{vmatrix}, \qquad D^{1/2}(I_y) = \frac{1}{2i}\begin{vmatrix} 0 & 1 \\ -1 & 0 \end{vmatrix}, \\
&D^{1/2}(I_x) = \frac{1}{2}\begin{vmatrix} 0 & 1 \\ 1 & 0 \end{vmatrix}, \\
&D^{1/2}(I_+) = \begin{vmatrix} 0 & 1 \\ 0 & 0 \end{vmatrix}, \qquad D^{1/2}(I_-) = \begin{vmatrix} 0 & 0 \\ 1 & 0 \end{vmatrix}, \\
&D^{1/2}(I_x^2) = D^{1/2}(I_y^2) = D^{1/2}(I_z^2) = \tfrac{1}{4}\mathbf{1}, \\
&D^{1/2}(I^2) = \tfrac{3}{4}\mathbf{1}.
\end{aligned} \tag{3.97}$$

It has been stated that a rotation can be built up from three infinitesimal rotations. A rotation characterized by the three Eulerian angles ϕ, θ, χ is represented by the following matrix derived from the matrices for the infinitesimal rotations:

$$D^{1/2}(R_{\phi,\theta,\chi}) = \begin{vmatrix} \exp[i(\chi+\phi)/2]\cos(\theta/2) & \exp[i(\chi-\phi)/2]\sin(\theta/2) \\ -\exp[-i(\chi-\phi)/2]\sin(\theta/2) & \exp[-i(\chi+\phi)/2]\cos(\theta/2) \end{vmatrix} \tag{3.98}$$

The matrix changes its sign each time 2π is added to χ or ϕ. As pointed out previously, the matrices of the $D^{(j)}$ representation with $j = \frac{1}{2}$ being half an odd integer are double-valued and form a double group. In the consideration of double groups, it is customary to simplify the notation, denoting the matrix associated with the operation R by R if the matrix is taken without a change of sign, and by $\overline{R}$ if it is taken with a change of sign.

For a system involving an electron, double-group symmetry should be considered. The representations of the double group may be regarded as direct-product representations $D^s \times D^{1/2}$; D^s is a representation of the double group that has $D^s(R) = D^s(\overline{R})$. The D^s's apply if the electron spin is overlooked.

Improper Rotations. Consider now rotation groups that contain improper rotations. An improper rotation consists of a rotation followed by a reflection. If the group does not contain inversion as an element, it can be treated by considering the isomorphic group of proper rotations.

A group containing the inversion element is a direct product of the inversion group E, Π and a group of proper rotations. The inversion group must now be considered. The element Π^2 is the identity element E:

$$D(\Pi^2) = D(E).$$

For single-valued representations, $D(E) = \mathbf{1}$ corresponds to two different $D(\Pi)$'s which are two different representations having even parity, $D(\Pi) = +\mathbf{1}$, and odd parity, $D(\Pi) = -\mathbf{1}$, respectively. For a double-valued representation, $D(E) = E, \overline{E} = \mathbf{1}, -\mathbf{1}$, which corresponds to four different $D(\Pi)$'s: $\mathbf{1}$, $-\mathbf{1}$, $i\mathbf{1}$, $-i\mathbf{1}$. For the spin that is intrinsic to an electron, $D(\Pi)$ is the same for all spinors. Results of significance involve products $u_i u_j^*$ of spinors. Since $D(\Pi)[D(\Pi)]^* = \mathbf{1}$ is independent of the choice of $D(\Pi)$, $D(\Pi) = \mathbf{1}$ is usually taken.

c. Time-Reversal Symmetry

Define a time-reversal operator T that consists of taking the complex conjugate, and replace t with $-t$. Operating on the Schrödinger wave equation with T, we get

$$0 = T\left[\left(H - i\hbar\frac{\partial}{\partial t}\right)\right]\Psi = TH\Psi - (-i\hbar)\frac{\partial}{\partial(-t)}T\Psi$$
$$= \left(THT^{-1} - i\hbar\frac{\partial}{\partial t}\right)T\Psi. \qquad (3.99)$$

The Hamiltonian operator H represents energy and is independent of time due to energy conservation. Although H contains linear momenta $\mathbf{p}$ and angular momenta $\mathbf{L}$ and $\mathbf{S}$, each of which is represented by a pure imaginary operator, it contains only even powers of the momenta, such as $\mathbf{p}^2$, $\mathbf{p}_i \cdot \mathbf{S}_i$, and $\mathbf{S}_i \cdot \mathbf{S}_j$. Therefore H is invariant to replacing every operator by its complex conjugate. It follows that H commutes with T, and

$$THT^{-1} = H. \qquad (3.100)$$

The time-reversal operation is then a symmetry transformation of the Hamiltonian, and $T\psi$ is a wavefunction of the system.

For an energy eigenfunction,

$$T\Psi = T[\varphi(\mathbf{r}_1,\mathbf{r}_2,\ldots,\boldsymbol{\sigma}_1,\boldsymbol{\sigma}_2,\ldots)e^{-iEt/\hbar}] = \varphi^* e^{-iEt/\hbar}, \qquad (3.101)$$

showing that φ and φ^* belong to the same energy E. If φ and φ^* are independent, the degeneracy of the energy level will be increased by the time-reversal symmetry.

Consider the complex conjugates of spinors. The relationship between u_+^*, u_-^* and u_+, u_- can be found from the transformation properties. Let R be the transformation corresponding to a rotation of the coordinate axes x, y, z. We have

$$Ru_+ = au_+ + bu_- \qquad \text{and} \qquad Ru_- = cu_+ + du_-,$$

where the coefficients a, b, c, d depend on the rotation R. The complex conjugate of the above equations is

$$Ru_+^* = a^*u_+^* + b^*u_-^* \qquad \text{and} \qquad Ru_-^* = c^*u_+^* + d^*u_-^*.$$

According to (3.98),

$$a^* = d, \qquad b^* = -c, \qquad c^* = -b, \qquad d^* = a.$$

Hence we have

$$R(u_-^*) = a(u_-^*) + b(-u_+^*) \qquad \text{and} \qquad R(-u_+^*) = c(u_-^*) + d(-u_+^*).$$

Since $u_-^*, -u_+^*$ transform like u_+, u_-, they must be proportional to u_+, u_-, and we may take

$$u_+^* = -u_-, \qquad u_-^* = u_+, \tag{3.102}$$

and get

$$Tu_+ = u_+^* = -u_-, \qquad Tu_- = u_-^* = u_+, \tag{3.103}$$

For a one-electron eigenfunction φ that is a product of a spin function and a function of $\mathbf{r}$, we get

$$T^2\varphi = -\varphi.$$

A ϕ that is a product of n one-electron φ's is referred to as many-electron wavefunction; for such a function we get

$$T^2\phi = -\phi \text{ if } n \text{ is odd}, \qquad T^2\phi = \phi \text{ if } n \text{ is even}. \tag{3.104}$$

Time-Reversal Symmetry and Degeneracy. Operating with T on a function $F = aF_1 + bF_2$, we get

$$TF = a^*TF_1 + b^*TF_2, \tag{3.105}$$

instead of the

$$TF = aTF_1 + bTF_2$$

given by the linear transformations we have been dealing with. The time-reversal operation cannot be incorporated into the group representation scheme considered previously. In general, such an operator is said to be antilinear. However, the time-reversal operation does conserve the magnitude of the function and the magnitude of products of functions. Such an operator is said to be antiunitary.

For a group containing antiunitary operators, the representation requires special consideration. For a group of unitary operators R, the matrices of a representation satisfy the condition

$$D(R_i)D(R_j) = D(R_iR_j).$$

When the group includes antiunitary operators A, such a relation is not appropriate for all the group elements. Instead, the matrices should satisfy the relations

$$\begin{aligned} D(R_i)D(R_j) &= D(R_iR_j), \qquad & D(R_i)D(A_j) &= D(R_iA_j), \\ D(A_i)D(A_j)^* &= D(A_iA_j), \qquad & D(A_i)D(R_j)^* &= D(A_iR_j). \end{aligned} \tag{3.106}$$

Such a set of matrices is called a corepresentation of the nonunitary group. The following result has been derived.

The irreducible representation D of a group without its antiunitary operators belongs to one of the following three cases:

(a) D can be transformed to real form. There is an additional degeneracy, and D occurs twice (there is no degeneracy if the number of electrons is even or the spins are neglected).
(b) D and D^* are equivalent but cannot be transformed to real form. There is no additional degeneracy (there is, if the number of electrons is even or the spins are neglected). (3.107)
(c) D and D^* are inequivalent. There is an additional degeneracy, and D and D^* occur together.

The case to which D belongs can easily be found by using the characters of the representation:

$$\sum_R \chi(R^2) = \begin{cases} \text{(a)} & h \\ \text{(b)} & -h\,, \\ \text{(c)} & 0 \end{cases} \tag{3.108}$$

where h is the order of the group.

For the consideration of the group of vector $\mathbf{k}$ in connection with electronic energy bands, it has been shown that the R in (3.108) is to be taken as an element in the space group that takes $\mathbf{k}$ into $-\mathbf{k}$, and h is the number of such elements. R^2 is an element of the group of vector $\mathbf{k}$, and $\chi(R^2)$ is the character in the irreducible representation of the group of vector $\mathbf{k}$.

Kramers' Theorem. This well-known theorem states that all energy levels of a system containing an odd number of electrons must be at least doubly degenerate, provided there is no externally applied magnetic field. The minimum of degeneracy comes from time-reversal symmetry. In an externally applied magnetic field, energy conservation of the system may hold and the Hamiltonian is consequently real. However, the Hamiltonian contains interactions of the magnetic field with magnetic moments of the system in their linear power. Hence the Hamiltonian is not invariant without reversing the direction of the external magnetic field. The theorem does not exclude an externally applied electric potential, since the interaction added to the Hamiltonian involves only the positions $\mathbf{r}$ of charges in the system.

d. Examples of Energy Bands

Irreducible representations have been worked out for various structures. Consider, for example, crystals of the diamond and zinc blende structures. The much-studied elements silicon and germanium have the diamond structure, and many common semiconductors such as the group III–V compounds have the zinc blende structure. Both structures have a cubic Bravais lattice, the face-centered cubic lattice. There are three kinds of rotation axes in the cubic lattice: a fourfold rotation axis from the center to the midpoint of a face of the cube, a threefold rotation axis from the center to a corner of the cube, and a twofold rotation axis from the center to the midpoint of an edge of the cube. A cubic Bravais lattice has the $m3m$ point group, which consists of 48 elements divided into 10 classes: E, $6C_4$, $8C_3$, $6C_2$, $3C_4^2$, $I \times Z$. The operation is given by the symbol, and the number of such operations is given by the numeral in front of the symbol. For example, C_4 denotes a right or left rotation of $2\pi/4$ about a fourfold rotation axis, C_4^2 is two such rotations in succession, I stands for inversion, Z denotes one of the five classes of proper rotations (including E), and $I \times Z$ denotes a rotation followed by an inversion.

The point-group symmetry of a structure may be less than that of its Bravais lattice, depending on the basis ions of a lattice point. A lattice point of the diamond structure consists of two identical ions so arranged that the structure has symmetry elements of inversion and fourfold screw axes. The structure has the $m3m$ point group. On the other hand, a lattice point of the zinc blende structure consists of two ions of different kinds, and consequently the symmetry is reduced. The zinc blende structure has the point group $\overline{4}3m$. In particular, there is no symmetry of pure inversion and there is no screw axis in the elements of the space group.

Figure 3.1 shows the Brillouin zone (the first Brillouin zone or the reduced Brillouin zone) of a face-centered cubic lattice. Shown also are the usual notations for the points and lines of some symmetry. Let a wavevector **k** be denoted by its terminal point. Tables 3.1 through 3.4 tabulate characters for the irreducible representations of the points Γ and X. Listed in the first column of each table are the kinds of elements of the double group, two for each kind of physical operation R. The common notation of the representation is given in the top row of each table. A representation that has $\chi(R) = \chi(\overline{R})$ for every R is a D^s representation, which may not be consistent with spin symmetry. Such representations can be considered if the

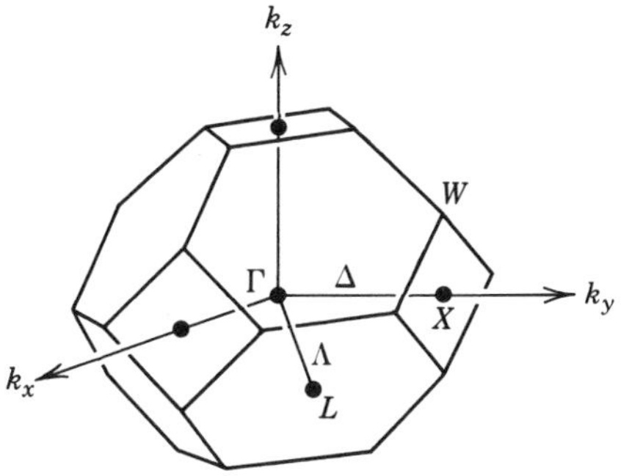

Figure 3.1. Brillouin zone of a face-centered cubic lattice.

spin can be overlooked. They are listed to the left of the vertical line. Listed to the right of the vertical line are the irreducible representations of the double group, which are proper for consideration in any case. Such a representation has $\chi(R) = -\chi(\overline{R})$.

Each table also gives $D^s \times D^{1/2}$ for each D^s. Shown are the compositions of $D^s \times D^{1/2}$ in terms of the irreducible representations of the double group. Correlation of the proper representations with the D^s representations provides a useful means for the investigation of the effect of spin.

Character tables can and have been constructed for any point. The evolution of representation over the Brillouin zone can then be obtained for an energy band by means of compatibility relations. For example, the compatibility relations given in Table 3.5 show the evolution of representation for the various types of energy bands as $\mathbf{k}$ varies along the line Γ–X.

To illustrate the effect of time-reversal symmetry, consider an arbitrary $\mathbf{k}$ that can be turned into $-\mathbf{k}$ only by inversion. In the zinc blende structure, which does not have inversion as a symmetry operation, there is no R in (3.108), and we have case (c) with additional degeneracy. For a Bloch function φ associated with $\mathbf{k}$, φ^* is associated with $-\mathbf{k}$. In making φ and φ^* correspond to the same energy $\epsilon(\mathbf{k}) = \epsilon(-\mathbf{k})$, the time-reversal symmetry adds degeneracy to the energy level ϵ. On the other hand, inversion is a symmetry operation of the diamond structure. Referring to (3.108), $R = I, \overline{I}$, and we have case (a) with additional degeneracy. In this case, $-\mathbf{k}$ is in the star of $\mathbf{k}$ and $\epsilon(\mathbf{k}) = \epsilon(-\mathbf{k})$ due to the inversion operation. The additional degeneracy given by the time-reversal symmetry pertains to the degeneracy of the energy level $\epsilon(\mathbf{k})$ for the wavevector $\mathbf{k}$.

TABLE 3.1 Character Table of the Double Group of Γ; Diamond Structure

	Γ_1	$\Gamma_{1'}$	Γ_2	$\Gamma_{2'}$	Γ_{12}	$\Gamma_{12'}$	Γ_{15}	$\Gamma_{15'}$	Γ_{25}	$\Gamma_{25'}$	$\Gamma_6^\pm$	$\Gamma_7^\pm$	$\Gamma_8^\pm$
E	1	1	1	1	2	2	3	3	3	3	2	2	4
$\overline{E}$	1	1	1	1	2	2	3	3	3	3	−2	−2	−4
$8C_3$	1	1	1	1	−1	−1	0	0	0	0	1	1	−1
$8\overline{C_3}$	1	1	1	1	−1	−1	0	0	0	0	−1	−1	1
$6C_2$	1	1	−1	−1	0	0	−1	−1	1	1	0	0	0
$6\overline{C_2}$	1	1	−1	−1	0	0	−1	−1	1	1	0	0	0
$6C_4$	1	1	−1	−1	1	1	1	1	−1	−1	$\sqrt{2}$	$-\sqrt{2}$	0
$6\overline{C_4}$	1	1	−1	−1	0	0	1	1	−1	−1	$-\sqrt{2}$	$\sqrt{2}$	0
$3C_4^2$	1	1	1	1	2	2	−1	−1	−1	−1	0	0	0
$3\overline{C_4^2}$	1	1	1	1	2	2	−1	−1	−1	−1	0	0	0
I	1	−1	1	−1	2	−2	−3	3	−3	3			
$\overline{I}$	1	−1	1	−1	2	−2	−3	3	−3	3			
$8I \times C_3$	1	−1	1	−1	−1	1	0	0	0	0			
$8\overline{I \times C_3}$	1	−1	1	−1	−1	1	0	0	0	0			
$6I \times C_2$	1	−1	−1	1	0	0	1	−1	−1	1	$\pm\chi(Z)$		
$6\overline{I \times C_2}$	1	−1	−1	1	0	0	1	−1	−1	1	$Z = E, \overline{E}, C_3,$		
$6I \times C_4$	1	−1	−1	1	0	0	−1	1	1	−1	$\overline{C_3}, \ldots, C$		
$6\overline{I \times C_4}$	1	−1	−1	1	0	0	−1	1	1	−1			
$3I \times C_4^2$	1	−1	1	−1	2	−2	1	−1	1	−1			
$3\overline{I \times C_4^2}$	1	−1	1	−1	2	−2	1	−1	1	−1			
D^s	Γ_1	$\Gamma_{1'}$	Γ_2	$\Gamma_{2'}$	Γ_{12}	$\Gamma_{12'}$	Γ_{15}	$\Gamma_{15'}$	Γ_{25}	$\Gamma_{25'}$			
$D^s \times D^{1/2}$	Γ_6^+	Γ_6^-	Γ_7^+	Γ_7^-	Γ_8^+	Γ_8^-	$\Gamma_6^- + \Gamma_8^-$	$\Gamma_6^+ + \Gamma_8^+$	$\Gamma_7^- + \Gamma_8^-$	$\Gamma_7^+ + \Gamma_8^+$			

e. Selection Rules for Matrix Elements

The integral

$$\int \varphi_i^* Q \varphi_j \, d\tau \tag{3.109}$$

is the matrix element of the operator Q between the states i and j, the integration being over all the variables. Various matrix elements are involved in theoretical treatments. Selection rules are the condition for a matrix element to be nonzero.

The integral of a function is invariant under a linear transforma-

TABLE 3.2 Character Table of the Double Group of Γ; Zinc Blende Structure

	Γ_1	Γ_2	Γ_3	Γ_4	Γ_5	Γ_6	Γ_7	Γ_8
E	1	1	2	3	3	2	2	4
$\overline{E}$	1	1	2	3	3	-2	-2	-4
$8C_3$	1	1	-1	0	0	1	1	-1
$8\overline{C}_3$	1	1	-1	0	0	-1	-1	1
$3I \times C_4^2$	1	1	2	-1	-1	0	0	0
$3\overline{I \times C_4^2}$	1	1	2	-1	-1	0	0	0
$6I \times C_4$	1	-1	0	-1	1	$\sqrt{2}$	$-\sqrt{2}$	0
$6\overline{I \times C_4}$	1	-1	0	-1	1	$-\sqrt{2}$	$\sqrt{2}$	0
$6I \times C_2$	1	-1	0	1	-1	0	0	0
$6\overline{I \times C_2}$	1	-1	0	1	-1	0	0	0
D^s	Γ_1	Γ_2	Γ_3	Γ_4	Γ_5			
$D^s \times D^{1/2}$	Γ_6	Γ_7	Γ_8	$\Gamma_7 + \Gamma_8$	$\Gamma_6 + \Gamma_8$			

tion T of the coordinates:

$$\int F\,d\tau = T\int F\,d\tau = \int (TF)\,d\tau. \tag{3.110}$$

Considering a group G of h elements of T, we have

$$h\int F\,d\tau = \sum_T \int (TF)\,d\tau. \tag{3.111}$$

An arbitrary function F may be expanded in terms of orthonormal functions of a complete set. The energy eigenfunctions of the system constitute such a set, and they are the basis functions of the irreducible representations of the Hamiltonian. Let $\varphi_i^{(\lambda r)}$ be the ith eigenfunction of the rth set that transforms according to the irreducible representation $D^{\lambda r}$. The expansion of F is

$$F = \sum_\lambda \sum_r \sum_i a_{\lambda r,i}\varphi_i^{(\lambda r)}, \tag{3.112}$$

where the a's are coefficients. Using the expansion, we get

$$\begin{aligned} h\int F\,d\tau &= \sum_\lambda \sum_r \sum_i a_{\lambda r,i} \sum_T D_{ji}^{\lambda r}(T) \int \varphi_j^{(\lambda r)}\,d\tau \\ &= \sum_\lambda \sum_r \sum_i a_{\lambda r,i} \left[\sum_T 1 \cdot D_{ji}^{\lambda r}(T)\right] \int \varphi_j^{(\lambda r)}\,d\tau \end{aligned} \tag{3.113}$$

TABLE 3.3 Character Table of the Double Group of X; Diamond Structure

	X_1	X_2	X_3	X_4	X_5
$(E0)$	2	2	2	2	4
$(\overline{E0})$	2	2	2	2	−4
(Ea)	2	2	2	2	4
$(\overline{Ea})$	2	2	2	2	−4
$(C_{4\parallel}^2 \mid 0)$	2	2	−2	−2	
$\overline{(C_{4\parallel}^2 \mid 0}$	2	2	−2	−2	
$2(C_{4\perp}^2 \mid 0)$	0	0	0	0	
$2\overline{(C_{4\perp}^2 \mid 0)}$	0	0	0	0	
$2(C_4 \mid \tau)$	0	0	0	0	
$2\overline{(C_4 \mid \tau)}$	0	0	0	0	
$2(C_2 \mid \tau)$	0	0	2	−2	
$2\overline{(C_2 \mid \tau)}$	0	0	2	−2	$\chi = 0$
$(I \mid \tau)$	0	0	0	0	
$\overline{(I \mid \tau)}$	0	0	0	0	
$(I \times C_{4\parallel}^2 \mid \tau)$	0	0	0	0	
$\overline{(I \times C_{4\parallel}^2 \mid \tau)}$	0	0	0	0	
$2(I \times C_{4\perp}^2 \mid \tau)$	0	0	0	0	
$2\overline{(I \times C_{4\perp}^2 \mid \tau)}$	0	0	0	0	
$2(I \times C_{4\parallel} \mid 0)$	0	0	0	0	
$2\overline{(I \times C_{4\parallel} \mid 0)}$	0	0	0	0	
$2(I \times C_2)$	2	−2	0	0	
$2\overline{(I \times C_2)}$	2	−2	0	0	

$X_i \times D^{1/2} = X_5,\ i = 1,2,3,4$

where the 1 in the square bracket denotes any one of the elements $D_{\alpha\beta}(T)$ of the identity representation. According to the orthogonality of irreducible representations (3.46), the term in square brackets vanishes unless $D^{\lambda r}$ is an identity representation.

The pertinent group to be considered for the matrix elements of a system is the symmetry group of the Hamiltonian of the system. First find the representation D of the group according to which the matrix element transforms. Then decompose the representation into irreducible representations. The matrix element is zero if D does not contain an identity representation.

TABLE 3.4 Character Table of the Double Group of X; Zinc Blende Structure

	X_1	X_2	$X_{3(x)}$	X_4	$X_{5(y,z)}$	X_6	X_7
E	1	1	1	1	2	2	2
$\overline{E}$	1	1	1	1	2	-2	-2
$2C_{4\perp}^2$	1	1	-1	-1	0	0	0
$\overline{2C_{4\perp}^2}$	1	1	-1	-1	0	0	0
$C_{4\parallel}^2$	1	1	1	1	-2	0	0
$\overline{C_{4\parallel}^2}$	1	1	1	1	-2	0	0
$2I \times C_{4\parallel}$	1	-1	-1	1	0	$\sqrt{2}$	$-\sqrt{2}$
$\overline{2I \times C_{4\parallel}}$	1	-1	-1	1	0	$-\sqrt{2}$	$\sqrt{2}$
$2I \times C_2$	1	-1	1	-1	0	0	0
$\overline{2I \times C_2}$	1	-1	1	-1	0	0	0
D^s	X_1	X_2	X_3	X_4	X_5		
$D^s \times D^{1/2}$	X_6	X_7	X_7	X_6	$X_6 + X_7$		

TABLE 3.5 Compatibility Relations Along the Line Δ

Diamond Structure		Zinc Blende Structure	
$\Gamma_1 \to \Delta_1$	$X_1 \to \Delta_1$	$\Gamma_1 \to \Delta_1$	$X_1 \to \Delta_1$
$\Gamma_{1'} \to \Delta_{1'}$	$X_1 \to \Delta_{1'}$	$\Gamma_2 \to \Delta_2$	$X_2 \to \Delta_2$
$\Gamma_2 \to \Delta_2$	$X_2 \to \Delta_{2'}$	$\Gamma_3 \to \Delta_1 + \Delta_2$	$X_3 \to \Delta_1$
$\Gamma_{2'} \to \Delta_{2'}$	$X_3 \to \Delta_{2'}$	$\Gamma_4 \to \Delta_1 + \Delta_3 + \Delta_4$	$X_4 \to \Delta_2$
$\Gamma_{12} \to \Delta_1 + \Delta_2$	$X_{3'} \to \Delta_2$	$\Delta_5 \to \Delta_2 + \Delta_3 + \Delta_4$	$X_5 \to \Delta_3 + \Delta_4$
$\Gamma_{12'} \to \Delta_{1'} + \Delta_{2'}$	$X_4 \to \Delta_{1'}$	$\Delta_6, \Gamma_7, \Gamma_8 \to \Delta_5$	$X_6, X_7 \to \Delta_5$
$\Gamma_{15} \to \Delta_1 + \Delta_5$	$X_{4'} \to \Delta_1$		
$\Gamma_{15'} \to \Delta_{1'} + \Delta_{5'}$	$X_5 \to \Delta_5$		
$\Gamma_{25} \to \Delta_2 + \Delta_5$	$X_{5'} \to \Delta_5$		
$\Gamma_{25'} \to \Delta_{2'} + \Delta_{5'}$			
$\Gamma_6 \to \Delta_6$			
$\Gamma_7 \to \Delta_7$			
$\Gamma_8 \to \Delta_6 + \Delta_7$			

f. Magnetic Crystals

For a crystal that spontaneously exhibits macroscopic magnetization, symmetry operations are subject to the invariance of magnetization. In general, the order of the symmetry point group is smaller, the primitive cell is larger, and the Brillouin zone is smaller, due to the existence of magnetization. Time reversal has a more important effect on symmetry. It has been pointed out that time reversal is a symmetry operation T of the Hamiltonian. Magnetization has a direction, and hence a perfect crystal exhibiting magnetization is not in a statistical equilibrium of energy eigenstates. Time reversal cannot be a symmetry for crystals in such a condition, since it reverses the direction of any magnetic moment. However, combinations TA_k of T with some unitary operators A_k may be symmetry operators. Thus both unitary and antiunitary operators are possible group elements of magnetic crystals. A theory of symmetry groups was worked out by Schubnikov in which an operation interchanging "black" and "white" colors was considered in addition to the geometric operations, the color change being analogous to the reversal of magnetic moment. The groups are referred to as Schubnikov groups or color groups.

There are 122 possible point groups and 1651 possible space groups for magnetic crystals compared to 32 and 230, respectively, for nonmagnetic crystals. Consider the magnetic point groups, of which there are three types. The 32 groups of the first type have no antiunitary operators, and each is the same as an ordinary point group. Crystals with groups of this type are either ferromagnetic or antiferromagnetic. For example, a ferromagnetic crystal with a magnetic point group of one rotation axis has the same point group that disregards the magnetization, an antiferromagnetic crystal with a magnetic point group of a twofold rotation axis and mirror planes parallel to it has the same point group that disregards the magnetization.

There are also 32 magnetic point groups of the second type. Each is formed by adding elements TA_i for every element A_i in the ordinary point group. This process doubles the number of symmetry elements and makes $T = TE$ an element. It has been pointed out that time reversal T cannot be a symmetry operation for a crystal having macroscopic magnetization. However, time reversal combined with a translation is not excluded. A symmetry operation including translation is represented in the point group by the operation minus translation, improper rotations being familiar operations of this kind. In the added elements TA_i, T actually stands for the time reversal combined with a primitive lattice translation of the crystal. Apparently, crystals having magnetic point groups of this type are antiferromagnetic or devoid of macroscopic magnetism, diamagnetic or paramagnetic.

There are 58 magnetic point groups of the last type. Each group contains time reversal only in combination with a spatial rotation or reflection, TA_k. The operators A_k cannot include those for which n

defined by $A_k^n = E$ is an odd integer, otherwise $(TA_k)^n = T^n(A_k)^n = T$ to be excluded would be an element of the group. Furthermore, the operators A_k must be distinct from the operators A_i that appear without T, since $(TA_k)A_k^{-1} = T$ would then be in the group. There are four more important properties concerning the structure of these groups.

1. For a group $G = \{A_i, TA_k\}$, the set $G^0 = \{A_i, A_k\}$ is also a group. It can easily be seen that the elements of G^0 have the same combination relations as the elements of group G.

2. The set $S = \{A_i\}$ of unitary operators is an invariant subgroup of G, that is,

$$(TA_k)S(TA_k)^{-1} = S \tag{3.114}$$

for every TA_k. The first step of the proof is given by the fact that T commutes with all spatial operators:

$$(TA_k)S(TA_k)^{-1} = T^2 A_k S A_k^{-1} = A_k S A_k^{-1}.$$

Second,

$$A_k S A_k^{-1} = S, \tag{3.115}$$

since the set $A_k S A_k^{-1}$ is a set of unitary operators of group G. It follows from the second step of proof that S is also an invariant subgroup of G^0.

3. The number of elements in G or G^0 is twice the number of elements in S. The proof is as follows. The multiplication of a TA_k by each A_i produces as many distinct TA_k as the number of A_i, showing the number of TA_k to be at least as large as the number of A_i. On the other hand, the multiplication of a TA_k by each TA_k produces as many distinct A_i as there are TA_k, showing the number of A_i to be at least as large as the number of TA_k. Consequently, the number of A_i must be equal to the number of TA_k.

4. If S is an invariant subgroup of index 2 in a point group G^0, then $G = S + T(G^0 - S)$ is a group, a magnetic point group of the type under consideration. To prove that G is a group, we need only show that

$$G \cdot G = G.$$

The elements $G^0 - S = C$ form a set of cosets of the group G^0 under the invariant subgroup S. It can be shown that $S \cdot S = S$, $S \cdot C = C \cdot S = C$, and $C \cdot C = S$. Hence

$$G \cdot G = (S + TC) \cdot (S + TC) = S + TC = G. \tag{3.116}$$

We now get a procedure for finding all 58 magnetic point groups of the above type. Take an ordinary point group G^0, find all its invariant subgroups S of index 2, and form the magnetic group $G = S + T(G^0 - S)$ for each S. Performing this process for all 32 groups G^0, one obtains all the magnetic groups G.

Irreducible representations. Although magnetic point groups are in structure analogous to color groups, they preclude ordinary matrix representation. On account of the antiunitary nature of the time-reversal operator T, shown by (3.105), corepresentation is required. For the effect of time reversal on degeneracy, the previously mentioned criteria (3.107) apply; the addition of degeneracy refers to subgroup S. In the consideration of electronic energy bands, it should be borne in mind that the Brillouin zone is not the same as in the absence of macroscopic magnetization; groups G^k and compatibility relations of representations are different.

CHAPTER FOUR
Electron States

A. ONE-ELECTRON STATES

1. ENERGY-BAND STATES

In the one-electron approximation, the exchange term in the Hartree–Fock equation complicates the calculation. It becomes difficult to determine the appropriate potential to use. Furthermore, in addition to the fact that the Fock Hamiltonian H^{F} in (3.5) does not adequately cover the correlation energy, the spin-dependent interactions given by (3.80) and (3.81) are not considered. It has not been possible to carry out calculations as satisfactorily as for a free atom. This section discusses a number of types of approximate calculations that have been made.

The Cellular Method. In this method, used by Wigner and Seitz for their calculations for alkali metals, the crystal is divided into cells, each cell covering the volume around one atom. The potential in a cell is taken to be the free-ion potential used in the calculation of atomic states. The electronic state $\mathbf{k} = 0$ is particularly suitable for the application of this method; by symmetry, $\varphi_{\mathbf{k}=0}(r)$ has zero gradient at the cell boundary in monatomic materials of simple structures. The calculation becomes analogous in the free-atom calculation except for a modification of the boundary condition, the modification being simplified if the cell is replaced by a sphere of the same volume. Calculations covering $\mathbf{k} \neq 0$ states have been made based on the cellular method; it appears that the problem of matching the wavefunctions at the cell boundary can be dealt with satisfactorily.

The Plane-Wave Method. The potential is assumed to be a superposition of free-atom potentials $v(\mathbf{r})$, each of which includes a free-electron exchange potential

$$V(\mathbf{r}) = \sum_n \sum_b v(\mathbf{r} - \mathbf{R}_{b,n}), \tag{4.1}$$

where

$$\mathbf{R}_{b,n} = \mathbf{a}_n + \mathbf{b}$$

is the position vector of ion b in the primitive cell n, $\mathbf{a}_n$ is the lattice vector of the primitive cell, and $\mathbf{b}$ is the position vector of ion b in the cell. Expanding the wavefunction in terms of normalized plane waves $|\mathbf{k}\rangle$ in a volume Ω,

$$\varphi(\mathbf{r}) = \sum_{\mathbf{k}} a_{\mathbf{k}} |\mathbf{k}\rangle = \Omega^{-1/2} \sum_{\mathbf{k}} a_{\mathbf{k}} \exp(i\mathbf{k} \cdot \mathbf{r}), \tag{4.2}$$

substituting the expanded wavefunction into the Schrödinger equation, multiplying the equation by $|\mathbf{k}'\rangle$, and integrating over $\mathbf{r}$, we get a set of equations

$$\frac{\hbar^2}{2m} k'^2 a_{\mathbf{k}'} + \sum_{\mathbf{k}} \langle \mathbf{k}' | V | \mathbf{k} \rangle = \epsilon a_{\mathbf{k}'}$$

for various $\mathbf{k}'$. The matrix element

$$\begin{aligned}\langle \mathbf{k}' | V | \mathbf{k} \rangle &= \frac{1}{\Omega} \int \exp(-i\mathbf{k}' \cdot \mathbf{r}) \sum_{n,b} v(\mathbf{r} - \mathbf{R}_{b,n}) \exp(i\mathbf{k} \cdot \mathbf{r})\, d\mathbf{r} \\ &= \frac{1}{\Omega} \left[\sum_b \exp[-i(\mathbf{k} - \mathbf{k}') \cdot \mathbf{b}] \int \exp[-i(\mathbf{k}' - \mathbf{k}) \cdot (\mathbf{r} - \mathbf{R}_{b,n})] \right. \\ &\quad \left. \times v(\mathbf{r} - \mathbf{R}_{b,n})\, d\mathbf{r} \right] \sum_n \exp[-i(\mathbf{k} - \mathbf{k}') \cdot \mathbf{a}_n] \end{aligned} \tag{4.3}$$

is equal to zero unless $\mathbf{k}' - \mathbf{k}$ is a reciprocal-lattice vector $\mathbf{G}$. This condition greatly reduces the number of simultaneous equations.

Furthermore, the matrix element decreases in magnitude with increasing $|\mathbf{k}' - \mathbf{k}|$. Despite these considerations, the number of simultaneous equations to be solved is very large in order to obtain a satisfactory solution for a set of expansion coefficients $a_{\mathbf{k}}$ that gives the energy and wavefunction of a state.

The Orthogonalized Plane-Wave Method. The wavefunctions of an electron oscillate near a nucleus as may be expected from wavefunctions in an atom. In the plane-wave expansion, many waves are needed to reproduce these oscillations. Orthogonalized plane waves (OPW) have been used in place of plane waves.

$$\mathrm{OPW}_{\mathbf{k}} = |\mathbf{k}\rangle - \sum_{t,i} |t,i\rangle\langle t,i \mid \mathbf{k}\rangle, \tag{4.4}$$

where t is the index of atomic core states and i is the index of atoms. $|t,i\rangle$ is the normalized core state $|t\rangle$ centered at ion i. In view of

$$\langle t,i \mid \mathbf{k}\rangle = \Omega^{-1/2} \int \psi_t^*(\mathbf{r} - \mathbf{R}_i) \exp(i\mathbf{k} \cdot \mathbf{r})\, d\mathbf{r},$$

$\mathrm{OPW}_{\mathbf{k}}$s are orthogonal to the core states, under the assumption that the core states of different ions do not overlap. The matrix elements coupling the OPWs decrease with increasing $|\mathbf{k}' - \mathbf{k}|$ much more rapidly than the matrix elements coupling the plane waves. A comparatively small number of $\mathrm{OPW}_{\mathbf{k}}$s may be adequate.

The Augmented Plane-Wave Method. In this method, the electron wavefunction is assumed to be a plane wave outside of certain spheres centered at the nuclei. Inside each sphere, eigenfunctions are calculated for a spherical potential that approximates the true potential. Combinations of the eigenfunctions are constructed such that each joins continuously with a plane wave at the boundaries of the spheres. Such an augmented plane wave (APW) is an approximate wavefunction of the electron. This method also involves less complex calculations than the plane-wave method.

The Pseudopotential Theory. The one-electron wavefunction expanded in terms of $\mathrm{OPW}_{\mathbf{k}}$ is

$$\varphi = \sum_{\mathbf{k}} a_{\mathbf{k}} \mathrm{OPW}_{\mathbf{k}} \equiv \sum_{\mathbf{k}} a_{\mathbf{k}} (1 - P)|\mathbf{k}\rangle = (1 - P) \sum_{\mathbf{k}} a_{\mathbf{k}} |\mathbf{k}\rangle \equiv (1 - P)\phi, \tag{4.5}$$

where

$$P = \sum_{t,i} |t,i\rangle\langle t,i|$$

is a projection operator. The function ϕ is the "pseudowavefunction." Substitution of φ into the one-electron Schrödinger equation gives

$$-\frac{\hbar^2}{2m}\nabla^2\phi + V(\mathbf{r})\phi - \left[-\frac{\hbar^2}{2m}\nabla^2 + V(\mathbf{r})\right]P\phi + \epsilon P\phi = \epsilon\phi, \tag{4.6}$$

which can be written as

$$-\frac{\hbar^2}{2m}\nabla^2\phi + W\phi = \epsilon\phi. \tag{4.7}$$

Assume that $V(\mathbf{r})$ is the same as that in the Schrödinger equation

$$\left[-\frac{\hbar^2}{2m}\nabla^2 + V(\mathbf{r})\right]\varphi_t = \epsilon_t\varphi_t \tag{4.8}$$

for the core states of each atom. Then

$$\left[-\frac{\hbar^2}{2m}\nabla^2 + V(\mathbf{r})\right]P = \sum_{t,i}\epsilon_{t,i}|t,i\rangle\langle t,i|, \tag{4.9a}$$

and

$$W = V(\mathbf{r}) + \sum_{t,i}(\epsilon - \epsilon_{t,i})|t,i\rangle\langle t,i|. \tag{4.9b}$$

W is called the pseudopotential. The pseudowavefunction is equal to the true wavefunction outside the cores, since P is zero there. Hopefully, it remains smooth in the core region, where the oscillation of φ may have been taken care of by the orthogonalization. There is thus some justification to expect W to be small such that approximate expressions may be used to obtain reasonable values for ϕ and ϵ. The true wavefunction can be calculated from ϕ.

The $\mathbf{k}\cdot\mathbf{p}$ Perturbation and Effective-Mass Method. The method has often been used to calculate electron states with wavevectors close to the wavevector $\mathbf{k}_0$ of an extremum of an energy band. In order to get the variation of electronic states more reliably, the spin-dependent interaction has been taken into account. However, interaction is added to a one-electron Hamiltonian instead of considered in the derivation of a one-electron Hamiltonian. The Hamiltonian used is

$$H = \frac{1}{2m}\mathbf{p}^2 + V + \frac{\hbar}{2m^2c^2}\mathbf{s} \times \nabla V \cdot \mathbf{p}, \tag{4.10}$$

the last term of which is the "spin–orbit interaction." Except for the omission of two terms of minor importance for our problem, this expression of H follows from Dirac's relativistic equation for an electron.

With the above Hamiltonian, the periodic part $u_{\mathbf{k}}$ of the Bloch function satisfies the equation

$$\left[\frac{1}{2m}(\mathbf{p}+\hbar\mathbf{k})^2 + V + \frac{\hbar}{2m^2c^2}\mathbf{s} \times \nabla V \cdot (\mathbf{p}+\hbar\mathbf{k})\right] u_{\mathbf{k}} = \epsilon_{\mathbf{k}} u_{\mathbf{k}}. \tag{4.11}$$

We note that $u_{\mathbf{k}}$ is a function of spatial coordinates $\mathbf{r}$ and spin coordinates $\boldsymbol{\sigma}$. For

$$\mathbf{k} = \mathbf{k}_0 + \Delta\mathbf{k},$$

we have

$$\Delta H = H_{\mathbf{k}} - H_{\mathbf{k}_0} = \frac{1}{2m}\left[\hbar^2(\Delta\mathbf{k})^2 + 2\hbar\mathbf{p}\cdot\Delta\mathbf{k} + 2\hbar^2\mathbf{k}_0\cdot\Delta\mathbf{k}\right] + \frac{\hbar^2}{2m^2c^2}\mathbf{s} \times \nabla V \cdot \Delta\mathbf{k}. \tag{4.12}$$

The functions u_{0n} of the states $\mathbf{k}_0$ in the various energy bands n form a complete set for the expansion of functions $u_{\mathbf{k}}$, which are periodic in the lattice.

$$\Delta\epsilon_{\mathbf{k}} = \epsilon_{\mathbf{k}} - \epsilon_{\mathbf{k}_0} \qquad \text{and} \qquad \Delta u_{\mathbf{k}} = u_{\mathbf{k}} - u_{\mathbf{k}_0} \tag{4.13}$$

may be obtained by treating ΔH as a perturbation. If the spin is overlooked and $k_0 = 0$ is assumed, then $\Delta\mathbf{k}$ is $\mathbf{k}$ and the ΔH consisting of two terms differs from that for free electrons only by $2\hbar\mathbf{p}\cdot\mathbf{k}$. Hence this approach is broadly referred to as the $\mathbf{k}\cdot\mathbf{p}$ perturbation method.

The first-order perturbation gives

$$\Delta u_{\mathbf{k}} = \sum_i \frac{\langle 0|\Delta H|0,i\rangle}{\epsilon_0 - \epsilon_{0,i}} u_{0,i}$$

and

$$(\Delta\epsilon_{\mathbf{k}})_1 = \langle 0|\Delta H|0\rangle. \tag{4.14}$$

The second-order perturbation gives

$$(\Delta\epsilon_{\mathbf{k}})_2 = \sum_i \frac{\langle 0|\Delta H|0,i\rangle\langle 0,i|\Delta H|0\rangle}{\epsilon_0 - \epsilon_{0,i}}. \tag{4.15}$$

0 denotes the state of $\mathbf{k}_0$ in the band under consideration. $0,i$ denotes the $\mathbf{k}_0$ state in another band of index i. Since $\mathbf{k}_0$ corresponds to an extremum of the energy band being considered, the terms of ΔH that are linear in $\Delta\mathbf{k}$ do not contribute to $(\Delta\epsilon_{\mathbf{k}})_1$. We get to the second-order perturbation

$$\epsilon_{\mathbf{k}} = \epsilon_{\mathbf{k}_0} + \frac{\hbar^2}{2m}(\Delta\mathbf{k})^2 + \frac{\hbar^2}{2m}(\Delta\mathbf{k})^2 \sum_i \frac{\langle 0|\mathbf{T}|0,i\rangle\langle 0,i|\mathbf{T}|0\rangle}{\epsilon_0 - \epsilon_{0,i}},$$

$$\mathbf{T} = \left(\frac{2}{m}\right)^{1/2}\left[\mathbf{p} + \hbar\mathbf{k}_0 + \frac{\hbar}{2mc^2}\mathbf{s}\times\Delta V\right]\cdot\Delta\mathbf{k}/|\Delta\mathbf{k}|. \tag{4.16}$$

The tensor of reciprocal effective mass, defined by

$$1/m^* \equiv \hbar^{-2}\nabla_{\mathbf{k}}\nabla_{\mathbf{k}}\epsilon = \hbar^{-2}\nabla_{\Delta\mathbf{k}}\nabla_{\Delta\mathbf{k}}\epsilon,$$

is given by

$$\frac{m}{m^*} = 1 + \sum_i \frac{\langle 0|\mathbf{T}|0,i\rangle\langle 0,i|\mathbf{T}|0\rangle}{\epsilon_0 - \epsilon_{0,i}}. \tag{4.17}$$

The quantity $\epsilon_{\mathbf{k}} - \epsilon_{\mathbf{k}_0}$ is equal to the eigenvalue of a Hamiltonian operator

$$H = \frac{\hbar^2}{2m}\left(\Delta\mathbf{k}\cdot\frac{m}{m^*}\cdot\Delta\mathbf{k}\right) \tag{4.18}$$

for a stationary wave of wavevector $\Delta\mathbf{k} = \mathbf{k} - \mathbf{k}_0$. The operator may be referred to as the Hamiltonian in the effective-mass approximation.

It should be noted that the treatment presented above applies to a nondegenerate extremum of the energy band. If the extremum is degenerate, such as the top of the valence band in many semiconductors, the calculation is more complicated.

2. LOCALIZED STATES

Imperfections in a crystal, physical defects or chemical impurities, introduce irregularities of potential in limited regions. The wavefunctions and the energy spectrum of an energy band are affected by the presence of an imperfection, much as the states of a free electron are affected by the presence of an ion. More important, an imperfection may introduce localized states, the wavefunctions of which diminish toward zero with increasing distance. An energy gap of the crystal would then have energy levels of localized states, or "localized energy levels." Such a level is called a donor level if the level with an electron in occupation does not alter the number of electrons in the energy bands and an acceptor level if it reduces that number by one.

The main difficulty in treating an imperfection is to find the appropriate potential. Consider the simplest case that has received much attention: A regular atom of the crystal is replaced by an atom of an element in a neighboring column of the periodic table. The impurity atom has one more or one less valence electron than the regular atom. Its effect may be considered to be similar to that of changing the ionic charge by $+e$ or $-e$ and adding or subtracting an electron, respectively. Clearly, the impurity atom can bind an electron in a localized state of a donor level or an electron deficiency or a hole in an acceptor level. Donor (acceptor) levels close to the edge of the nearest higher-lying (lower-lying) energy band are referred to as shallow levels. The localized wavefunction of such a level $\psi(\mathbf{r})$ should be extensive, covering a large number of lattice cells. It may be approximated by a linear combination of Bloch functions of the particular energy band in the neighborhood of the band edge:

$$\psi(\mathbf{r}) = \sum_{\mathbf{k}} A_{\mathbf{k}} \varphi_{\mathbf{k}}(\mathbf{r}). \tag{4.19}$$

It can be shown that the Fourier transform $F(\mathbf{r})$ of the expansion coefficients $A_{\mathbf{k}}$,

$$F(\mathbf{r}) = V^{-1/2} \sum_{\mathbf{k}} A_{\mathbf{k}} \exp[i(\mathbf{k} \cdot \mathbf{r})], \tag{4.20}$$

is a solution of the equation

$$[\epsilon_{\mathbf{k}}(-i\nabla) + V(\mathbf{r})]F(\mathbf{r}) = EF(\mathbf{r}). \tag{4.21}$$

The operator $\epsilon_{\mathbf{k}}(-i\nabla)$ is $\epsilon(\mathbf{k})$ of the energy band with $\mathbf{k}$ replaced by $-i\nabla$. The potential $V(\mathbf{r})$ is that of a point charge in a medium having the static dielectric constant ε_0 of the crystal:

$$V(\mathbf{r}) = -e^2/\varepsilon_0 r.$$

In the approximation $\epsilon(\mathbf{k}) \propto k^2$, the equation for $F(\mathbf{r})$ resembles the Schrödinger equation for a hydrogen atom, and hence the solutions obtained are often referred to as hydrogenic. We note that $F(\mathbf{r})$ gives the envelope of the localized wavefunction. Solutions of $F(\mathbf{r})$ that do not approach zero with increasing r belong to the energy band, in analogy with the wavefunctions of the ionized electron of a hydrogen atom.

The outlined treatment has to be corrected for the difference between a point charge and the impurity ion. The so-called core correction is the more important the more concentrated the envelope function. Elaborate treatments have to be made if $F(\mathbf{r})$ turns out to be confined to a few unit cells around the impurity ion. For example, an impurity atom of a group V element introduces donor states in the energy gap of the semiconductor silicon. According to the hydrogenic approximation, the ground state of each impurity atom is ~ 29 meV from the edge of the conduction band, whereas the experimental values, 44 meV for phosphorus, 49 meV for arsenic, 39 meV for antimony, and 69 meV for bismuth, are different from each other and significantly different from the value given by the simple approximation.

3. ELECTRONIC STATES UNDER AN APPLIED MAGNETIC FIELD

Consider the eigenstates of conduction carriers. The effect of an applied magnetic field has been treated rigorously by Landau for free electrons. The treatment can be applied to conduction carriers in the effective-mass approximation. The Hamiltonian of motion under a magnetic field $\mathbf{B}$ of vector potential $\mathbf{A}$ is then

$$H = \frac{1}{2m}\left(\mathbf{p} - \frac{e}{c}\mathbf{A}\right)\frac{m}{m^*}\left(\mathbf{p} - \frac{e}{c}\mathbf{A}\right). \tag{4.22}$$

In a material medium,

$$\nabla \times \mathbf{A} = \mathbf{B}.$$

In the usual case of interest, the permeability μ is a scalar. We take

$$A_x = A_z = 0 \qquad \text{and} \qquad A_y = Bx \tag{4.23}$$

for a uniform $B = B_z$. The plane-wave part, $\exp(i\mathbf{k} \cdot \mathbf{r})$, of the Bloch function is replaced by a function $\chi(\mathbf{r})$. Consider first the case of a scalar m^*. The equation for $\chi(\mathbf{r})$ is

$$-\frac{\hbar^2}{2m^*}\left[\frac{\partial^2}{\partial z^2} + \frac{\partial^2}{\partial z^2} + \left(\frac{\partial}{\partial y} + \frac{ie}{\hbar c}Bx\right)^2\right]\chi = \chi. \tag{4.24}$$

The equation is satisfied by $\chi(\mathbf{r})$ of the following form:

$$\chi(\mathbf{r}) = \exp[i(k_y y + k_z z)]F(x). \tag{4.25}$$

The equation to be satisfied by $F(x)$ is

$$\left[-\frac{\hbar^2}{2m^*}\frac{d^2}{dx^2} + \frac{m^*}{2}\left(\frac{eB}{m^*c}\right)^2\left(x - \frac{\hbar c}{eB}k_y\right)^2\right]F(\mathbf{x})$$
$$= \left(\epsilon - \frac{\hbar^2}{2m^*}k_z^2\right)F(\mathbf{x}). \tag{4.26}$$

This is the equation of a harmonic oscillator with mass m^*, a natural frequency $\omega_c = |eB|/m^*c$, and a center at $(\hbar c/eB)k_y$. The eigenvalues of the oscillator energy are

$$\epsilon_n = \frac{\hbar^2}{2m^*}k_z^2 = (n + \tfrac{1}{2})\hbar\omega_c, \tag{4.27}$$

where n is an integer. ω_c is called the cyclotron frequency, and the energy levels quantized by the discreteness of n are referred to as Landau levels.

There are states with various k_y and various k_x that have the same energy of specified n and k_z. According to the Born–von Karman boundary condition, the allowable values of k_y are multiples of $2\pi/L_y$, where L_y is the dimension of the specimen in the y direction. Furthermore, the center of $F(x)$, $(\hbar c/eB)k_y$, should be inside the specimen, a condition that requires that $|(k_y)_{\max} - (k_y)_{\min}| = (|eB|/\hbar c)L_x$ for a length L_x of the specimen in the x direction. Hence the number of states for a given k_z of a Landau level, a value of n, is

$$\frac{(|eB|/\hbar c)L_x}{2\pi/L_y} = \frac{|eB|}{2\pi\hbar c}L_x L_y.$$

Since two adjacent levels are separated by $\hbar\omega_c$ in energy, the number of states of the same k_z per unit energy is

$$\frac{|eB|}{2\pi\hbar c}L_x L_y \frac{1}{\hbar\omega_c} = \frac{m^*}{2\pi\hbar^2}L_x L_y.$$

This number is the same as that in the absence of magnetic field. The effect of the magnetic field is to quantize the energy with regard to the motion normal to $\mathbf{B}$ without changing the total number of states. The diagram of ϵ versus k_z is a series of equally spaced parabola, each of which is a Landau subband corresponding to a value of n. Each section Δk_z of a subband contributes

$$\frac{|eB|}{2\pi\hbar c}L_xL_y\frac{L_z}{2\pi}\Delta k_z \tag{4.28}$$

to the number of states of one spin. The number of states per unit energy per unit volume, the density of states, of a subband n is

$$\rho_n = \frac{|eB|}{4\pi^2\hbar c}\left(\frac{\partial\epsilon_n}{\partial k_z}\right)^{-1} = \frac{|eB|m^{*1/2}}{4\sqrt{2}\pi^2\hbar^2 c}\epsilon_n^{-1/2}, \tag{4.29}$$

where

$$\epsilon_n = \frac{\hbar^2}{2m^*}k_z^2 = \epsilon - (n+\tfrac{1}{2})\hbar\omega_c. \tag{4.30}$$

ρ_n increases with decreasing ϵ_n or $|k_z|$, approaching infinity as ϵ_n approaches zero. The total density of states $\rho(\epsilon) = \sum_n \rho_n$ has sharp peaks at energies $\epsilon = (n+\frac{1}{2})\hbar\omega_c$.

Consider now the case of a tensorial effective mass. Let $(1,2,3)$ denote the coordinate system in which the tensor $1/m^*$ is diagonalized with $m_1 = \mu_1 m$, $m_2 = \mu_2 m$, and $m_3 = \mu_3 m$. In the coordinate system $(1,2,3)$, the magnetic field may be characterized by three mutually perpendicular unit vectors $\hat{a}$, $\hat{b}\,\|\,\mathbf{B}$, $\hat{c}\,\|\,\mathbf{A}$, the direction cosines of which can be used to transform coordinates from $(1,2,3)$ to a system with an axis along $\mathbf{B}$ and an axis along $\mathbf{A}$. Reduce the scale of $(1,2,3)$ along each of the axes i by the corresponding $(\mu_i)^{1/2}$. Then rotate the reduced system to a coordinate system (x,y,z), in which the resulting $A_x = A_z = 0$ and A_y does not involve z. The three unit vectors for the coordinate transformation from $(1,2,3)$ to (x,y,z) are

$$\hat{\alpha} \perp \hat{b}/\mu^{1/2}, \quad \hat{\beta}\,\|\,\hat{b}/\mu^{1/2}, \quad \text{and} \quad \hat{\gamma}\perp\hat{b}/\mu^{1/2}, \quad \hat{\gamma}\perp\hat{a}/\mu^{1/2}, \tag{4.31}$$

where $\hat{a}/\mu^{1/2}$, for example, stands for a vector with components $a_1/\mu_1^{1/2}$, $a_2/\mu_2^{1/2}$, and $a_3/\mu_3^{1/2}$. In the coordinate system (x,y,z), the Schrödinger equation becomes

$$\frac{\hbar^2}{2m}\left[\frac{\partial^2}{\partial x^2} + \frac{\partial^2}{\partial z^2} + \left(\frac{\partial}{\partial y} - \frac{ieB}{\hbar c\mu_c}x - \frac{ieB}{\hbar cM}y\right)^2\right]\chi = \epsilon\chi. \tag{4.32}$$

The solution is

$$\chi = \exp\left[i\left(k_z z + k_y y + \frac{eB}{2\hbar cM}y^2\right)\right]F(x)$$

and

$$\epsilon = \frac{\hbar^2 k_z^2}{2m} + (n+\tfrac{1}{2})\frac{\hbar|eB|}{cm\mu_c} = \frac{\hbar^2 [k_z]^2}{2m\mu_z} + (n+\tfrac{1}{2})\frac{\hbar|eB|}{cm\mu_c}, \qquad (4.33)$$

where

$$\mu_c = \left(\frac{\mu_1\mu_2\mu_3}{\mu_1 c_1^2 + \mu_2 c_2^2 + \mu_3 c_3^2}\right)^{1/2}, \qquad \mu_z = \frac{\mu_1^2 c_1^2 + \mu_2^2 c_2^2 + \mu_3^2 c_3^2}{\mu_1 c_1^2 + \mu_2 c_2^2 + \mu_3 c_3^2},$$

$$M = \left(\frac{a_1 b_1}{\mu_1} + \frac{a_2 b_2}{\mu_2} + \frac{a_3 b_3}{\mu_3}\right)^{-1},$$

and $[k_z]$ is the wavevector in actual rather than reduced units. The density of states of a Landau subband, per unit actual volume, is given by

$$\rho_n = \frac{|eB|(\mu_d m)^{1/2}}{4\sqrt{2}\pi^2\hbar^2 c}\epsilon_n^{-1/2}, \qquad (4.34)$$

where

$$\mu_d = c_1^2\mu_1 + c_2^2\mu_2 + c_3^2\mu_3.$$

Magnetic Flux. $\mu(x^2+y^2)$ is an area in the xy plane. Consider for simplicity the case of a scalar effective mass. The function $\chi(\mathbf{r})$ given by (4.25) is normalized as follows:

$$\int_0^1 dz \int_0^1 dy \int_{-\infty}^{+\infty} \chi^* \chi \, dx = 1.$$

For the function χ, the expectation value of the area perpendicular to the magnetic field $\mathbf{B} \parallel \hat{z}$ is

$$\begin{aligned}\pi \int \chi_n^*(x^2+y^2)\chi_n \, d\mathbf{r} &= \pi \int_{-\infty}^{+\infty} F_n^*(x) x^2 F_n(x)\, dx + \frac{\pi}{3} \\ &= \pi(n+\tfrac{1}{2})\frac{\hbar c}{|e|B} + \frac{\pi}{3}, \qquad (4.35)\end{aligned}$$

where n is the index of the Landau subband. This area A and the magnetic flux $\Phi = BA$ through this area are quantized among the subbands. Evidently, an analogous quantization occurs if the effective mass is a tensor.

The Effect of Spin. A particle of charge e has a magnetic moment represented by the operator $2(e\hbar/2mc)\mathbf{s} = 2\beta\mathbf{s}$, where $\mathbf{s}$ is the operator of spin angular momentum and β is the Bohr magneton.

Under an applied field **B**, the magnetic moment of spin contributes a term,

$$-2\beta\mathbf{s}\cdot\mathbf{B}, \tag{4.36}$$

to the Hamiltonian of the particle. Another, often more important, effect of the spin appears in the part of the Hamiltonian (4.10) through spin–orbit coupling. Replacing the operator $\mathbf{p}$ with $\mathbf{p} - e\mathbf{A}/c$ adds a term $(\hbar^2/2m^2c^2)\mathbf{s}\times\nabla V\cdot(-e\mathbf{A}/c)$, which involves the magnetic field. It can be shown that, as a consequence, this part of the Hamiltonian becomes

$$\frac{\hbar^2}{2m}\left(\mathbf{p}-\frac{e}{c}\mathbf{A}\right)\cdot\frac{m}{m^*}\cdot\left(\mathbf{p}-\frac{e}{c}\mathbf{A}\right)+\frac{1}{im}\sum_i{}'\frac{\pi_{0i}\times\pi_{i0}}{\epsilon_0-\epsilon_i}\beta\cdot\mathbf{B}. \tag{4.37}$$

The subscript i is the index of the energy band, and the subscript 0 indicates the band being considered. m/m^* is the tensor (4.17), which is independent of the magnetic field, and the operator π is

$$\pi = \mathbf{p} + \frac{\hbar}{2mc^2}\mathbf{s}\times\nabla V. \tag{4.38}$$

The total one-electron Hamiltonian under an applied magnetic field may be written as

$$H = \frac{\hbar^2}{2m}\left(\mathbf{p}-\frac{e}{c}\mathbf{A}\right)\cdot\frac{m}{m^*}\cdot\left(\mathbf{p}-\frac{e}{c}\mathbf{A}\right) - \beta\mathbf{s}\cdot g\cdot\mathbf{B} \tag{4.39}$$

with

$$\mathbf{s}\cdot g = 2\mathbf{s} + \frac{1}{im}\sum_i{}'\frac{\pi_{0i}\times\pi_{i0}}{\epsilon_0-\epsilon_i}.$$

The tensor g is called the g factor. The second term on the right in the above expression may be predominant; without it the g factor would have the value 2. The magnitude of this term is large when the magnitude of m/m^* is large. For example, the conduction band of InSb with a large $m/m^* \approx 70$ has a $g = -50$.

The energy band extremum being considered is doubly degenerate, with spin taken into account. Each of the two states is represented by a two-component spinor. Under the applied magnetic field **B**, each Landau subband is split into two branches, due to the second term of the Hamiltonian. The energy splitting is given by the difference between the eigenvalues of this term for the two branches. The density of states of each band is given by (4.34).

B. MANY-ELECTRON STATES

1. DIELECTRIC SCREENING

In connection with correlation energy, we have pointed out the inadequacy of the one-electron approximation in regard to electron–electron interaction. The screening of electric field in a many-electron system is one of the problems. Another has to do with the formation of stationary states not given by the one-electron approximation.

Consider the Fourier expansions of an existing electric field $\mathbf{E}$ and the associated displacement $\mathbf{D}$:

$$\mathbf{E} = \sum_{\kappa} \mathbf{E}_{\kappa} \exp(i\boldsymbol{\kappa}\cdot\mathbf{r}), \qquad \mathbf{D} = \sum_{\kappa} \exp(i\boldsymbol{\kappa}\cdot\mathbf{r}). \tag{4.40}$$

The dielectric function $\varepsilon(\omega,\boldsymbol{\kappa})$ is

$$\varepsilon(\omega,\boldsymbol{\kappa}) = \mathbf{D}_{\kappa}/\mathbf{E}_{\kappa} = 1 + 4\pi(\mathbf{P}_{\kappa}/\mathbf{E}_{\kappa}), \tag{4.41}$$

where ω is the frequency of the field and $\mathbf{P}_{\kappa}$ is the $\boldsymbol{\kappa}$ component of the polarization in the system. (If the electrons have a background medium, the polarization of the background has to be taken into consideration.) The dielectric function is referred to as longitudinal if $\mathbf{E}_{\kappa} \parallel \boldsymbol{\kappa}$ and transverse if $\mathbf{E}_{\kappa} \perp \boldsymbol{\kappa}$. It is a scalar quantity if $\mathbf{D} \parallel \mathbf{E}$.

Consider the introduction of a test charge $\delta\rho$ with density distribution

$$\delta\rho = \delta\rho_{\kappa} \exp[i(\boldsymbol{\kappa}\cdot\mathbf{r} - \omega t)] + \text{c.c.} \tag{4.42}$$

According to the relation

$$\operatorname{div}\mathbf{D} = 4\kappa(\text{test charge density}), \tag{4.43}$$

we get

$$i\boldsymbol{\kappa}\cdot\mathbf{D}_{\kappa} = 4\pi\delta\rho_{\kappa} e^{-i\omega t} \qquad \text{and} \qquad \mathbf{D}_{\kappa} \parallel \boldsymbol{\kappa} \tag{4.44}$$

by equating terms of the same $\boldsymbol{\kappa}$ on the two sides of the relation. The electric field $\mathbf{E}$ produced by the test charge is given by

$$\operatorname{div}\mathbf{E} = 4\pi(\text{test charge density+induced charge density}). \tag{4.45}$$

Let $\Delta\rho = \Delta\rho_{\kappa} \exp(i\boldsymbol{\kappa}\cdot\mathbf{r}) + \Delta\rho_{-\kappa} \exp(-i\boldsymbol{\kappa}\cdot\mathbf{r})$ denote the induced charge density. We have

$$i\boldsymbol{\kappa}\cdot\mathbf{E}_{\kappa} = 4\pi(\delta\rho_{\kappa} e^{-i\omega t} + \langle\Delta\rho_{\kappa}\rangle). \tag{4.46}$$

Using (4.44) and (4.46), we get from (4.41)

$$\frac{1}{\varepsilon(\omega,\boldsymbol{\kappa})}=1+\frac{\langle\Delta\rho_{\kappa}\rangle}{\delta\rho_{\kappa}e^{-i\omega t}}. \tag{4.47}$$

For sufficiently small ω such that the vector potential of the oscillating charge may be neglected, we have Coulomb interaction between the system and the test charge:

$$H_1=(4\pi/\kappa^2)\rho_{-\kappa}\delta\rho_{\kappa}e^{-i\omega t+st}+\text{c.c.}, \tag{4.48}$$

where s is a small positive quantity introduced to represent adiabatic switching of the interaction at some $t<0$. In cases where the many-electron system may be regarded as isotropic, $\mathbf{E}_{\kappa}\parallel\mathbf{D}_{\kappa}$, which shows that the dielectric function is a scalar and that it is a longitudinal dielectric function in view of $\mathbf{E}_{\kappa}\parallel\boldsymbol{\kappa}$.

Consider the interaction H_1 given by (4.48) as a perturbation of the system. Let ψ_m denote the eigenfunctions of the unperturbed system, with a subscript indicating the eigenstate. To the first order in perturbation, the wavefunction Ψ of the system is given by

$$\Psi=\psi_0+\sum_m{}'\frac{4\pi}{\kappa^2}\delta\rho_{\kappa}\left(\frac{\langle m|\rho_{-\kappa}|0\rangle\exp[-i(\omega-\omega_{m0}+is)t]}{\omega-\omega_{m0}+is}\right.$$
$$\left.+\frac{\langle m|\rho_{\kappa}|0\rangle\exp[i(\omega+\omega_{m0}+is)t]}{-(\omega+\omega_{m0}+is)}\right)\psi_m, \tag{4.49}$$

where $\hbar\omega_{m0}=\epsilon_m-\epsilon_0$. We note that for a state m, $\langle m|\rho_{\kappa}|0\rangle$ and $\langle m|\rho_{-\kappa}|0\rangle$ cannot both be nonzero. We get, to the first order of $\delta\rho_{\kappa}$,

$$\langle\Delta\rho_{\kappa}\rangle=\langle\psi|\rho_{\kappa}|\psi\rangle-\langle\psi_0|\rho_{\kappa}|\psi_0\rangle=\langle\psi|\rho_{\kappa}|\psi\rangle$$
$$=\frac{4\pi}{\kappa^2}\delta\rho_{\kappa}e^{-i(\omega+is)t}\sum_m\left[\frac{|\langle m|\rho_{-\kappa}|0\rangle|^2}{\omega-\omega_{m0}+is}+\frac{|\langle m|\rho_{\kappa}|0\rangle|^2}{-(\omega+\omega_{m0}+is)}\right]. \tag{4.50}$$

The spectrum of states connected with $|0\rangle$ by ρ_{κ} is identical with that connected by $\rho_{-\kappa}$. Furthermore, the matrix elements squared are correspondingly equal. Substitution of expression (4.50) for $\Delta\rho_{\kappa}$ in (4.47) gives

$$\frac{1}{\varepsilon(\omega,\boldsymbol{\kappa})}=1+\frac{4\pi}{\kappa^2}\sum|\langle m|\rho_{-\kappa}|0\rangle|^2[(\omega-\omega_{m0}+is)^{-1}$$
$$-(\omega+\omega_{m0}+is)^{-1}]. \tag{4.51}$$

For an isotropic system, this expression gives the scalar longitudinal dielectric function.

The dielectric function may also be treated by considering the system in the presence of a scalar potential:

$$\varphi_{\kappa} \exp[i(\kappa \cdot \mathbf{r} - \omega t)] + \text{c.c.} \tag{4.52}$$

Assuming the vector potential to be negligible, we have the associated electric field

$$\mathbf{E}_{\kappa} = -i\kappa\varphi_{\kappa} \exp[i(\kappa \cdot \mathbf{r} - \omega t)]. \tag{4.53}$$

The potential field adds a term to the one-electron Hamiltonian:

$$e\varphi_{\kappa} \exp[i(\kappa \cdot \mathbf{r} - \omega t)].$$

The term is treated as a perturbation. In the one-electron approximation, we get from first-order perturbation

$$\langle \Delta\rho_{\kappa} \rangle = e^2 \varphi_{\kappa} \sum_{\mathbf{k}} \frac{f_0(\mathbf{k}+\kappa) - f_0(\mathbf{k})}{\epsilon_{\mathbf{k}+\kappa} - \epsilon_{\kappa} + \omega + is}, \tag{4.54}$$

where $f_0(\mathbf{k})$ is the unperturbed distribution function of electrons and $\mathbf{k}$ is the wavevector of the one-electron state. The polarization $\mathbf{P}$ associated with $\langle \Delta\rho_{\kappa} \rangle$ is given by

$$\langle \Delta\rho_{\kappa} \rangle = \nabla \cdot \mathbf{P}_{\kappa} = i\kappa \cdot \mathbf{P}_{\kappa}. \tag{4.55}$$

The dielectric function is then given by

$$\varepsilon(\omega, \kappa) = 1 + 4\pi \left(\frac{\mathbf{P}_{\kappa}}{\mathbf{E}_{\kappa}} \right) = 1 - \frac{4\pi e^2}{\kappa^2} \left[\sum_{\mathbf{k}} \frac{f_0(\mathbf{k}+\kappa) - f_0(\mathbf{k})}{\epsilon_{\mathbf{k}+\kappa} - \epsilon_{\mathbf{k}} + \omega + is} \right]_{s \to +0}. \tag{4.56}$$

It is a longitudinal dielectric function since it is one for $\mathbf{E}_{\kappa} \parallel \kappa$ as indicated by (4.53). For an isotropic system, $\langle \Delta\rho_{\kappa} \rangle$ and $\mathbf{P}_{\kappa}$ are independent of the direction of κ, and hence the dielectric function is a scalar.

Consider the simple case where the electron energy is $\epsilon = \hbar^2 k^2 / 2m$. The electron gas is degenerate, having a Fermi energy ϵ_{F} with corresponding wavevectors of magnitude k_{F}. The "dielectric constant" $\varepsilon(0, \kappa)$ given by (4.56) for such a case is

$$\varepsilon(0, \kappa) = 1 + \frac{6\pi n e^2 / \epsilon_{\mathrm{F}}}{2\kappa^2} \left(\frac{k_{\mathrm{F}}}{\kappa} - \frac{\kappa}{4k_{\mathrm{F}}} \right) \ln \left| \frac{\kappa + 2k_{\mathrm{F}}}{\kappa - 2k_{\mathrm{F}}} \right|, \tag{4.57}$$

where n is the electron concentration. A singularity occurs at $\kappa = 2k_{\mathrm{F}}$. For $\kappa < 2k_{\mathrm{F}}$, ε is relatively large as some $\mathbf{k}$'s can be found which give small values of $\epsilon_{\mathbf{k}+\kappa} - \epsilon_k$ in the denominator of (4.56). On the other hand, no denominators are small for $\kappa > 2k_{\mathrm{F}}$. It has been pointed out that the sharp decrease of ε as κ increases through $2k_F$ should lead to a sharp increase of the phonon energy at $\kappa = 2k_F$, since the stiffness of the lattice is increased by the enhancement of electron response to the ion motion.

For κ small compared to k_F, (4.57) reduces to

$$\varepsilon(0,\boldsymbol{\kappa}) = (\kappa^2 + 6\pi n e^2/\epsilon_F)/\kappa^2. \tag{4.58}$$

This is the well-known Thomas–Fermi expression for the dielectric constant of a degenerate electron gas.

2. EXCITONS

In the one-electron approximation, the wavefunction of a system is taken to be a determinant Φ of single-particle eigenfunctions φ. Two states of the system may differ from each other by having one different φ. Actually, such a modification of Φ does not fully represent the change of state of the system if the electron–electron interaction is taken into account more accurately. This effect is significant and easily detected experimentally in the excitation of one electron across the energy gap of an insulator or semiconductor. For the eigenstates of the system to be dealt with, called excitons, it may be justifiable to consider the effect of electron interaction only for the two energy bands, which we refer to as electron–hole interaction for brevity.

The wavefunction of an exciton may be expressed in the form

$$\psi = \sum_k \sum_{k'} A_{c\mathbf{k},v\mathbf{k}'} \Phi_{c\mathbf{k},v\mathbf{k}'}, \tag{4.59}$$

where c and v are indices of the conduction and valence bands, respectively, and $\Phi_{c\mathbf{k},v\mathbf{k}'}$ denotes a determinant for the two bands, in which a Bloch function $\varphi_{v\mathbf{k}'}$ is replaced by $\varphi_{c\mathbf{k}}$. In the absence of electron–hole interaction, each $\Phi_{c\mathbf{k},v\mathbf{k}'}$ would give an excited state of the crystal. The problem is to determine the coefficients $A_{c\mathbf{k},v\mathbf{k}'}$ so as to diagonalize the Hamiltonian containing the interaction. The translational symmetry of the crystal requires that ψ be multiplied only by a constant, under translation by a lattice vector $\mathbf{R}$. On the other hand, each $\Phi_{c\mathbf{k},v\mathbf{k}'}$ is multiplied by a particular factor $\exp[i(\mathbf{k}-\mathbf{k}')\cdot\mathbf{R}]$. Therefore, a ψ should correspond to a definite

$\mathbf{K} \equiv \mathbf{k} - \mathbf{k}'$:

$$\psi_{\mathbf{K}} = \sum_{\mathbf{k}} A_{c\mathbf{k},v(\mathbf{k}-\mathbf{K})} \Phi_{c\mathbf{k},v(\mathbf{k}-\mathbf{K})}$$
$$= N^{-1/2} \sum_{\mathbf{k}} \Phi_{c\mathbf{k},v(\mathbf{k}-\mathbf{K})} \sum_{\mathbf{R}} \left[G_{\mathbf{K}}(\mathbf{R}) \exp\left(\frac{i\mathbf{K}\cdot\mathbf{R}}{2}\right)\right] \exp(-i\mathbf{k}\cdot\mathbf{R}), \tag{4.60}$$

where

$$G_{\mathbf{K}}(\mathbf{R}) \exp\left(\frac{i\mathbf{K}\cdot\mathbf{R}}{2}\right) = N^{-1/2} \sum_{\mathbf{k}} A_{\mathbf{k},(\mathbf{k}-\mathbf{K})} \exp(i\mathbf{k}\cdot\mathbf{R})$$

and N is the number of lattice cells.

The following equation for $G_{\mathbf{K}}(\mathbf{R})$, derived by Wannier, diagonalizes the Hamiltonian approximately:

$$\left[\epsilon_c(-i\nabla + \tfrac{1}{2}\mathbf{K}) - \epsilon_v(-i\nabla - \tfrac{1}{2}\mathbf{K}) - V(\mathbf{R})\right] G_{\mathbf{K}}(\mathbf{R}) = E G_{\mathbf{K}}(R). \tag{4.61}$$

The operators $\epsilon_c(-i\nabla + \frac{1}{2}\mathbf{K})$ and $\epsilon_v(-i\nabla - \frac{1}{2}\mathbf{K})$ are derived from the one-electron energies $\epsilon_c(\mathbf{k})$ and $\epsilon_v(\mathbf{k})$, respectively. The function $G_{\mathbf{K}}(\mathbf{R})$ is to be considered a function of a continuous variable $\mathbf{R}$. The normalized solution $G_{\mathbf{K}}(\mathbf{R})$ of this equation should be multiplied by $\Omega^{1/2}$ to give the coefficients for discrete lattice vectors $\mathbf{R}$, Ω being the volume of the unit cell. The equation is a kind of two-particle wave equation. The potential energy $V(\mathbf{R})$ consists of matrix elements of $e^2/(\mathbf{r}_i - \mathbf{r}_j)$ between various Bloch functions. For large values of R, $V(\mathbf{R})$ approaches a Coulomb potential screened by the static dielectric constant ε of the crystal:

$$V(\mathbf{R}) \to -e^2/\varepsilon R. \tag{4.62}$$

The appropriate $V(\mathbf{R})$ to use for different situations has been the subject of many discussions. We note that certain terms of the Hamiltonian are neglected in the derivation of the equation. It has been pointed out that more refined considerations may split some degenerate states given by the equation and make significant the orientation of the exciton wavefunction relative to the wavevector $\mathbf{K}$.

Various other treatments of the problem have been made for excitons of strong binding, that is, for which the minimum energy of exciton production is considerably smaller than the energy gap between the two bands. To some extent these treatments adopt the atomic point of view. The so-called excitation models treat the problem of an electron in the potential field of a positive hole located on an ion, using some combination of wavefunctions of the ion and its neighbors. The "electron-transfer model" treats the transfer of an electron from one ion to another ion or to a group of other ions; wavefunctions of various excited states of each ion are combined to form the wavefunction of the electron. Models of both types have been applied to alkali halides. The "Frenkel exciton" is an extreme model of tight binding. The exciton is considered a local phenomenon confined essentially to one ion. This model has been used for molecular crystals and rare-gas solids.

Effective Mass Theory of Excitons. Consider excitons of weak binding, for which the Wannier equation is applicable. For energies close to the ground state of the exciton, the important one-electron states are those of energy ϵ close to the bottom of the conduction band or close to the top of the valence band. The effective-mass approximation may therefore be used for ϵ_c and ϵ_v. Consider the simple case where the effective masses m_c and m_v are each a scalar. Taking $\max(\epsilon_v) = 0$, we have

$$\epsilon_v(\mathbf{k}) = -\frac{\hbar^2}{2m_v}(\mathbf{k}-\mathbf{k}_{v0})^2 \qquad \text{and} \qquad \epsilon_c(\mathbf{k}) = \epsilon_g + \frac{\hbar^2}{2m_c}(\mathbf{k}-\mathbf{k}_{c0})^2, \tag{4.63}$$

where ϵ_g is the energy gap and $\mathbf{k}_{v0}$, $\mathbf{k}_{c0}$ are the wavevectors of the top of the valence band and the bottom of the conduction band, respectively. The Wannier equation leads to

$$\left[\frac{\hbar^2}{2m_c}\left(-i\nabla + \frac{\mathbf{K}_1}{2}\right)^2 + \frac{\hbar^2}{2m_v}\left(-i\nabla - \frac{\mathbf{K}_1}{2}\right)^2 + V(\mathbf{R})\right] H_{\mathbf{K}}(\mathbf{R}) = (E-\epsilon_g)H_{\mathbf{K}}(\mathbf{R}), \tag{4.64}$$

in which

$$\mathbf{K}_1 = \mathbf{K} - (\mathbf{k}_{c0} - \mathbf{k}_{v0}) = (\mathbf{k}-\mathbf{k}_{c0}) - [(\mathbf{k}-\mathbf{K}) - \mathbf{k}_{v0}] \tag{4.65}$$

and

$$H_{\mathbf{K}}(\mathbf{R}) = \exp[-i(\mathbf{k}_{c0} + \mathbf{k}_{v0}) \cdot \mathbf{R}/2] G_{\mathbf{K}}(\mathbf{R}).$$

The equation can further be written in the following form:

$$\left[-\frac{\hbar^2}{2\mu}\nabla^2 + V(\mathbf{R})\right] F_{\mathbf{K}}(\mathbf{R}) = \left[E - \epsilon_g + \frac{\hbar^2 K_1^2}{2(m_c+m_v)}\right] F_{\mathbf{K}}(\mathbf{R}), \tag{4.66}$$

where

$$F_{\mathbf{K}}(\mathbf{R}) = \exp(-i\alpha\mathbf{K}\cdot\mathbf{R})G_{\mathbf{K}}(\mathbf{R}), \qquad \alpha = \frac{1}{2}\frac{m_c - m_v}{m_c + m_v} \tag{4.67}$$

and

$$1/\mu = 1/m_c + 1/m_v.$$

Using expression (4.62) for $V(\mathbf{R})$, we get an equation similar to that of a hydrogen atom with the following energy eigenvalues for the exciton:

$$E = \epsilon_g - \frac{e^4\mu}{2\hbar^2\varepsilon^2}\frac{1}{n^2} + \frac{\hbar^2 K_1^2}{2(m_c + m_v)}, \tag{4.68}$$

where n is an integer quantum number. The first two terms represent the energy associated with the relative motion of two particles having a reduced mass. The last term represents the kinetic energy associated with the motion of the center of mass of the two particles, and it leads to an energy band for each discrete n. An exciton state with $E \gtrsim \epsilon_g$ approaches a state of a free electron with a free hole.

Besides the wavevector $\mathbf{K}$, an exciton state is also characterized by the quantum number n, which is indicated by a superscript (n). A pair of electron–hole states, $c\mathbf{k}$ and $v(\mathbf{k}-\mathbf{K})$, are contained in the exciton state $\psi_{\mathbf{K}}^{(n)}$ with a weighting factor $A^{(n)}_{c\mathbf{k},v(\mathbf{k}-\mathbf{K})}$. From (4.60), (4.65), and (4.67), we get

$$A^{(n)}_{c\mathbf{k},v(\mathbf{k}-\mathbf{K})} = \left(\frac{\Omega}{N}\right)^{1/2} \sum_{\mathbf{R}} \exp\left[i\left(-\mathbf{k} + \frac{\mathbf{K}}{2} + \alpha\mathbf{K}_1\right)\cdot\mathbf{R}\right] F_{\mathbf{K}}^{(n)}(\mathbf{R}). \tag{4.69}$$

Exciton states of low energy E are of primary interest, since they differ most clearly from states of a free electron with a free hole. Consider then the case of $\mathbf{K}_1 = 0$. The associated $\mathbf{K}$ is $\mathbf{k}_{c0} = \mathbf{k}_{v0}$, and the exponent in (4.69) is

$$-\mathbf{k} + \frac{\mathbf{K}}{2} = -(\mathbf{k} - \mathbf{k}_{c0}) - \frac{\mathbf{k}_{c0} + \mathbf{k}_{v0}}{2} = -[(\mathbf{k} - \mathbf{K}) - \mathbf{k}_{v0}] - \frac{\mathbf{k}_{c0} + \mathbf{k}_{v0}}{2}. \tag{4.70}$$

If the orbit of electron–hole relative motion as indicated by $F(\mathbf{R})$ is large compared to the atomic distance, the coefficient $A_{c\mathbf{k},v(\mathbf{k}-\mathbf{K})}$ is comparatively large for small values of $\mathbf{k} - \mathbf{k}_{c0}$ and $[(\mathbf{k} - \mathbf{K}) - \mathbf{k}_{v0}]$. In other words, the one-electron states of the conduction and valence bands close to the energy gap are important for the exciton.

It should be pointed out that an energy band having an energy extremum at $\mathbf{k} \neq 0$ has similar extrema at symmetrically equivalent $\mathbf{k}$'s. Each peak of the valence band forms exciton states with every valley of the conduction band. The problem is somewhat complicated if the effective mass at the peak and/or the valley is a tensor. It is more complicated if the peak and/or the valley consists of degenerate bands.

Biexcitons. A biexciton refers to a state that, in the one-electron approximation, results from replacing two valence-band states j with two conduction-band states i. The equation corresponding to (4.61) has the form of an equation for four particles, with a potential V that is a function of six variables: $\mathbf{R}_{ii'}$, $\mathbf{R}_{jj'}$, $\mathbf{R}_{ij}$, $\mathbf{R}_{ij'}$, $\mathbf{R}_{i'j}$, $\mathbf{R}_{i'j'}$. The biexciton is often referred to as an exciton molecule. In the effective-mass approximation, the equation corresponding to (4.64) resembles the equation for a unit of two hydrogen atoms. Biexciton states have been revealed experimentally for a number of semiconductors.

Electron–Hole Liquid (EHL). At a sufficient concentration of electrons and holes, an electron–hole liquid may form that is distinct from a gas of excitons or biexcitons. In a number of semiconductors, the existence of an electron–hole liquid is indicated in experimental observations of various properties, the luminescence spectrum in particular. Theoretical treatments have been made, using various approximations of electron–electron interaction among the electrons and holes. The concentration of electrons and holes in the liquid estimated by the treatments is fairly consistent with the experimental deductions. The concentration depends on the temperature and differs considerably for different materials. In general, it is of the order of magnitude of 10^{18} cm^{-3}, which is three to four orders higher than the necessary concentration for the formation of biexcitons.

3. PLASMONS

Consider a system of electrons with a uniform background of positive charge. The electrons are assumed to interact only electrostatically. The Hamiltonian of the electrons in a unit volume is then simply

$$H_0 = \sum_{i=1}^{n} \frac{p_i^2}{2m}, \tag{4.71}$$

where i is the index of an electron and n is the concentration of electrons. On the average, the electrostatic interaction of electrons contributes an energy

$$H_1 = \frac{1}{2} \int \delta\rho \varphi(\mathbf{r}) d^3\mathbf{r}, \tag{4.72}$$

where $\delta\rho$ is the variation of charge density and $\varphi(\mathbf{r})$ is the average electrostatic potential given by $\delta\rho$. $\varphi(\mathbf{r})$ and $\delta\rho$ are related by the Poisson equation

$$\nabla^2 \varphi = -4\pi \delta\rho. \tag{4.73}$$

Expand the vector-displacement operator $\mathbf{q}$:

$$\mathbf{q} = n^{-1/2} \sum_{\mathbf{k}} \mathbf{Q_k} \exp(i\mathbf{k} \cdot \mathbf{r}) \tag{4.74}$$

where $\mathbf{Q_k} \parallel \mathbf{k}$, and $\mathbf{k}$ is a reduced vector for a unit volume. For the collective motion of electrons, the variation of charge density is

$$\delta\rho = en \operatorname{div} \mathbf{q} = ien\left(n^{-1/2} \sum_{\mathbf{k}} kQ_{\mathbf{k}} \exp(i\mathbf{k} \cdot \mathbf{r})\right). \tag{4.75}$$

Expanding the potential

$$\varphi = \sum_{\mathbf{k}} \varphi_{\mathbf{k}} \exp(i\mathbf{k} \cdot \mathbf{r}), \tag{4.76}$$

we get the Poisson equation in the form

$$\sum_{\mathbf{k}} k^2 \varphi_{\mathbf{k}} \exp(i\mathbf{k} \cdot \mathbf{r}) = i4\pi en\left[n^{-1/2} \sum_{\mathbf{k}} kQ_{\mathbf{k}} \exp(i\mathbf{k} \cdot \mathbf{r})\right].$$

It follows that

$$\varphi_{\mathbf{k}} = i4\pi en\left[n^{-1/2} (Q_{\mathbf{k}}/k) \exp(i\mathbf{k} \cdot \mathbf{r})\right]. \tag{4.77}$$

Substituting the expressions of $\delta\rho$ and φ into the interaction Hamiltonian H_1, we get

$$\begin{aligned} H_1 &= \frac{1}{2} \int d^3\mathbf{r}\left(ien^{1/2} \sum_{\mathbf{k}} kQ_{\mathbf{k}} \exp(i\mathbf{k} \cdot \mathbf{r})\right) \\ &\quad \times \left(i4\pi en^{1/2} \sum_{\mathbf{k}} k^{-1} Q_{\mathbf{k}} \exp(i\mathbf{k} \cdot \mathbf{r})\right) \\ &= 2\pi e^2 n \sum_{\mathbf{k}} \sum_{\mathbf{k}'} Q_{\mathbf{k}} Q_{\mathbf{k}'} \int d^3\mathbf{r} \exp[i(\mathbf{k} + \mathbf{k}') \cdot \mathbf{r}] \\ &= 2\pi e^2 n \sum_{\mathbf{k}} Q_{\mathbf{k}} Q_{-\mathbf{k}}. \end{aligned} \tag{4.78}$$

Now we expand the momentum operator $\mathbf{p}$,

$$\mathbf{p} = n^{-1/2} \sum_{\mathbf{k}} \mathbf{P}_{\mathbf{k}} \exp(-i\mathbf{k} \cdot \mathbf{r}), \tag{4.79}$$

and thereby express H_0 in the following form:

$$\begin{aligned} H_0 &= \frac{1}{2mn} \sum_i \left(\sum_{\mathbf{k}} \mathbf{P}_{\mathbf{k}} \exp(-i\mathbf{k} \cdot \mathbf{r}_i) \right)^2 \\ &= \frac{1}{2mn} \sum_{\mathbf{k}} \sum_{\mathbf{k}'} \mathbf{P}_{\mathbf{k}} \cdot \mathbf{P}_{\mathbf{k}'} \sum_i \exp[-i(\mathbf{k} + \mathbf{k}') \cdot \mathbf{r}_i] \\ &= \frac{1}{2m} \sum_{\mathbf{k}} \mathbf{P}_{\mathbf{k}} \cdot \mathbf{P}_{-\mathbf{k}}. \end{aligned} \tag{4.80}$$

The Hamiltonian of the electrostatically interacting electrons is then

$$H = H_0 + H_1 = \sum_{\mathbf{k}} \left(\frac{1}{2m} \mathbf{P}_{\mathbf{k}} \cdot \mathbf{P}_{-\mathbf{k}} + 2\pi e_n^2 Q_{\mathbf{k}} Q_{-\mathbf{k}} \right). \tag{4.81}$$

The operators $\mathbf{Q}_{\mathbf{k}}$ and $\mathbf{P}_{\mathbf{k}}$ have the commutation relation

$$\begin{aligned} [\mathbf{Q}_{\mathbf{k}}, \mathbf{P}_{\mathbf{k}'}] &= \frac{1}{n} \left[\sum_i \mathbf{q}_i \exp(+i\mathbf{k} \cdot \mathbf{r}_i), \sum_j \mathbf{p}_j \exp(-i\mathbf{k}' \cdot \mathbf{r}_j) \right] \\ &= \frac{1}{n} \sum_i \sum_j [\mathbf{q}_i, \mathbf{p}_j] \exp[i(\mathbf{k} \cdot \mathbf{r} - \mathbf{k}' \cdot \mathbf{r})] \\ &= \frac{1}{n} i\hbar n \delta_{\mathbf{k}\mathbf{k}'} = i\hbar \delta_{\mathbf{k}\mathbf{k}'}, \end{aligned}$$

which shows that $\mathbf{Q}_{\mathbf{k}}$ and $\mathbf{P}_{\mathbf{k}}$ are canonically conjugate operators. Hence the Hamiltonian is equivalent to that of a group of harmonic oscillators, each of which is characterized by an energy quantum

$$\hbar\omega_{\mathrm{p}} = \hbar(4\pi n e^2/m). \tag{4.82}$$

$\hbar\omega_{\mathrm{p}}$ is referred to as the plasmon. The energy eigenvalues of the system are multiples of the plasmon. In the above discussion, the electrons or charge carriers are assumed to have the same mass. For conduction electrons or conduction holes near the edge of an energy band, the effective-mass approximation holds and the m in ω_{p} is the effective mass.

4. SPIN DENSITY WAVE (SDW) AND CHARGE DENSITY WAVE (CDW)

The Hartree–Fock equation (3.4) with Hamiltonian operator H^{F} is the basis of electron states in one-electron approximation. For a periodic lattice of positive ions, the one-electron wavefunctions are Bloch functions, on the assumption that H^{F} has the translational periodicity of the lattice, and the charge density of electrons per unit lattice cell is a constant for the ground state of the crystalline substance. However, H^{F} contains the interaction between electrons of parallel spins. Consider the group of conduction electrons in a metal. Suppose the spin-up and spin-down electrons each have a spatially varying density:

$$\begin{aligned} \rho_+(\mathbf{r}) &= \tfrac{1}{2}\rho_0[1 + A\cos(\mathbf{Q}\cdot\mathbf{r} + \varphi)], \\ \rho_-(\mathbf{r}) &= \tfrac{1}{2}\rho_0[1 + A\cos(\mathbf{Q}\cdot\mathbf{r} - \varphi)]. \end{aligned} \tag{4.83}$$

The case of $\varphi = \pi/2$ is referred to as a spin density wave (SDW), which has a spatial variation in the total density of electron spins but no variation in the total density of electrons. The case of $\varphi = 0$ is referred to as a charge density wave (CDW), which is just opposite to SDW with regard to the variations of electron spin density and electron density. It can be shown that such a spatial variation reduces the exchange energy of the electrons; the reduction is the same for a SDW and a CDW of the same $\mathbf{Q}$, since the exchange interaction applies only to electrons of parallel spins.

It has been pointed out in Section III.1 that the Hartree–Fock approximation leaves out the interaction between electrons of antiparallel spins, which gives the correlation energy. It has been shown that the correlation energy is positive for a SDW and negative for a CDW. As a result, a SDW may increase the energy of conduction electrons, and it is significant only for metals of suitable characteristics. Chromium has been found to be a metal of this kind.

For a CDW, the energy of conduction electrons is reduced by the combination of exchange and correlation energies. However, a wave of charge density produces a wave of electric field that tends to inhibit the existence of the CDW. If the lattice of positive ions has no elastic rigidity at all, then the ions will be displaced correspondingly so that an electric field does not occur. Therefore the existence of a CDW ground state depends on the softness of the ion lattice. Experimental studies at very low temperatures have given evidence for the existence of CDWs in several metals.

The parameters A and $\mathbf{Q}$ of a SDW or a CDW that occurs in a metal are determined by minimization of the free energy of the substance. The lattice of ions is, of course, determined according to the same consideration, and its difference from the lattice for $\rho_+ = \rho_- =$ constant is sometimes referred to as Peierl's instability. The importance of lattice distortion is evident for a CDW, as discussed in the preceding paragraph. The parameter $\mathbf{Q}$ is generally incommensurate with the undistorted reciprocal lattice. It is easy to see that the following simple consideration is significant for the CDW in a perfectly deformable lattice. In this case, only the exchange energy and correlation energy of the conduction electrons need be considered. A wave of electron density with wavevector $\mathbf{Q}$ is associated with the presence of a wave component of $\mathbf{Q}$ in the one-electron Hamiltonian. This component produces a perturbation of the one-electron states, introducing an energy gap at the electron wavevector $\mathbf{k} = \pm\mathbf{Q}$. The energy is increased for states above the gap and decreased for states below the gap, the magnitude of energy change being bigger for states closer to the gap. Therefore the occurrence of CDW, the most reduction of energy, is optimized if the Fermi energy is inside the energy gap, that is, if

$$\mathbf{Q} = 2\mathbf{k}_F, \tag{4.84}$$

where $\mathbf{k}_F$ is a wavevector of the Fermi surface.

5. QUASIPARTICLES

It is preferable to begin the treatment of a many-body system by taking into consideration the interactions that are important for the phenomenon of interest. As a result of the interactions, basic elements of the system may collectively form groups that have generally come to be called quasiparticles. Excitons in an insulator or semiconductor and plasmons in the case of high carrier concentration are energy quanta of electron groups, quasiparticles resulting from electron–electron interaction. Polarons (Section V.1.d) are energy quanta of a quasiparticle consisting of an electron and associated lattice vibration. In the theory of superconductivity outlined in Chapter VI, Section VI.E, Cooper pairs may be considered quasiparticles.

In a similar sense, a phonon is the energy quantum of a normal mode, a quasiparticle of lattice vibration. Polaritons, presented in Section VI.C.2.e for an ionic crystal, are associated with quasiparticles consisting of lattice displacement and electromagnetic waves. Finally, magnons, Section VI.D.4.b, represent magnetic energy of spin waves, or quasiparticles of a lattice of spins.

CHAPTER FIVE

Electron–Lattice Interaction

The system of electrons has been considered under the assumption that the ions in the lattice stay fixed at their equilibrium positions. Actually, the electronic state is affected by the motion of ions or the positions of ions in the adiabatic approximation. As mentioned previously, the effect is treated as a perturbation.

1. INTERACTION OF ELECTRONS WITH LATTICE VIBRATION

In the one-electron approximation, the interaction introduces a perturbation potential, which may be expressed as

$$\Delta V = \sum_{Lb} \Delta \mathbf{R}_{Lb} \cdot \mathbf{W}_{Lb}(\mathbf{r}), \tag{5.1}$$

where $\mathbf{r}$ is the coordinate of an electron and $\mathbf{R}_{Lb}$ is the coordinate of ion b in the cell L of the lattice. $\Delta\mathbf{R}_{Lb}$ is given by (2.33). The matrix element of ΔV connecting two states of the crystal has the following expression:

$$\langle D', n_{\mathbf{q}p} \pm 1 | \Delta V | D, n_{\mathbf{q}p} \rangle = N^{-1/2} \sum_{\mathbf{q}p} \sum_{b} \left(\frac{\hbar}{2M_b \omega_{\mathbf{q}b}} \right)^{1/2}$$
$$\times \begin{pmatrix} (n_{\mathbf{q}p}+1)^{1/2} \\ n_{\mathbf{q}p}^{1/2} \end{pmatrix} \sum_{L} \exp(\mp i\mathbf{q} \cdot \mathbf{L}) \langle D' | \mathbf{W}_{Lb}(\mathbf{r}) \cdot \begin{pmatrix} e_{\mathbf{q}pb} \\ e^{*}_{\mathbf{q}pb} \end{pmatrix} | D \rangle. \tag{5.2}$$

The lattice part of the system is specified by the quantum number $n_{\mathbf{q}p}$ of the vibrational mode $\mathbf{q}p$ involved. The electronic part is specified by D. Since $\Delta\mathbf{R}_{Lb}$ is linear in the operators a^+ and a, ΔV connects states that differ by one quantum of a single mode of vibration.

It follows from the periodicity of the lattice that

$$\mathbf{W}_{Lb}(\mathbf{r}) = \mathbf{W}_{0b}(\mathbf{r} - \mathbf{L}). \tag{5.3}$$

Furthermore, $\mathbf{W}(\mathbf{r})$ is a function of the coordinate $\mathbf{r}$ of one electron, and the electronic system is represented by a determinant of one-electron wavefunctions $\varphi_{\mathbf{k}}$. Therefore,

$$\begin{aligned}\langle D'|\mathbf{W}_{Lb}(\mathbf{r})|D\rangle = \langle \mathbf{k}'|\mathbf{W}_{Lb}(\mathbf{r})|\mathbf{k}\rangle &= \int \varphi^*_{\mathbf{k}'}(\mathbf{r})\mathbf{W}_{0b}(\mathbf{r}-\mathbf{L})\varphi_{\mathbf{k}}(\mathbf{r})\,d\mathbf{r} \\ &= \int \varphi^*_{\mathbf{k}'}(\mathbf{r}'+\mathbf{L})\mathbf{W}_{0b}(\mathbf{r}')\varphi_{\mathbf{k}}(\mathbf{r}'+\mathbf{L})\,d\mathbf{r}' \\ &= \exp[i(\mathbf{k}-\mathbf{k}')\cdot\mathbf{L}]\int \varphi^*_{\mathbf{k}'}(\mathbf{r}')\mathbf{W}_{0b}(\mathbf{r}')\varphi_{\mathbf{k}}(\mathbf{r}')\,d\mathbf{r}' \\ &\equiv \exp[i(\mathbf{k}-\mathbf{k}')\cdot\mathbf{L}](\mathbf{I}_b/N),\end{aligned} \tag{5.4}$$

where $\varphi_{\mathbf{k}}$ is normalized for N lattice cells. The sum

$$\begin{aligned}&\sum_L \exp(\pm i\mathbf{q}\cdot\mathbf{L})\langle \mathbf{k}'|\mathbf{W}_{Lb}(\mathbf{r})\cdot\begin{pmatrix} e_{\mathbf{q}pb} \\ e^*_{\mathbf{q}pb}\end{pmatrix}|\mathbf{k}\rangle \\ &\qquad = \frac{\mathbf{I}_b}{N}\cdot\begin{pmatrix} e_{\mathbf{q}pb} \\ e^*_{\mathbf{q}pb}\end{pmatrix}\sum_L \exp[i(\mathbf{k}-\mathbf{k}'\pm\mathbf{q})\cdot\mathbf{L}]\end{aligned} \tag{5.5}$$

does not vanish only when $(\mathbf{k}-\mathbf{k}'\pm\mathbf{q})$ in the exponent is a reciprocal-lattice vector $\mathbf{G}$. It is equal to

$$\mathbf{I}_b\cdot\mathbf{e}_{\mathbf{q}pb} \qquad \text{for} \quad \mathbf{k}' = \mathbf{k}-\mathbf{q}+\mathbf{G},$$

which belongs to a phonon emission process given by the creation operator a^+. It is equal to

$$\mathbf{I}_b\cdot\mathbf{e}^*_{\mathbf{q}pb} \qquad \text{for} \quad \mathbf{k}' = \mathbf{k}+\mathbf{q}+\mathbf{G},$$

which belongs to a phonon absorption process due to the annihilation operator a. A process with $\mathbf{G}=0$ is called a normal, or N, process, and a process with $\mathbf{G}\neq 0$ is called an umklapp, or U, process. The matrix element of the perturbation potential can now be written as

$$\langle \mathbf{k} \mp \mathbf{q} + \mathbf{G}, n_{\mathbf{q}p} \pm 1 | \Delta V | \mathbf{k}, n_{\mathbf{q}p} \rangle = N^{-1/2} \sum_{\mathbf{q}p} \sum_{b} \left(\frac{\hbar}{2M_b \omega_{\mathbf{q}p}} \right)^{1/2} \mathbf{I}_b \begin{cases} \mathbf{e}_{\mathbf{q}pb} (n_{\mathbf{q}p} + 1)^{1/2} \\ \mathbf{e}^*_{\mathbf{q}pb} n^{1/2}_{\mathbf{q}p} . \end{cases} \tag{5.6}$$

The conditions indicated above regarding **k**, **k**′, and **q** are referred to as wavevector conservation. For a transition caused by the perturbation to occur, the energy of the system has to be conserved, which requires

$$\epsilon(\mathbf{k}') = \epsilon(\mathbf{k}) - \hbar\omega_{\mathbf{q}p} \qquad \text{for phonon emission}$$

and

$$\epsilon(\mathbf{k}') = \epsilon(\mathbf{k}) + \hbar\omega_{\mathbf{q}p} \qquad \text{for phonon absorption.} \tag{5.7}$$

Various approximations have been used for the electron–lattice interaction potential. The rigid-ion theory assumes that the potential U contributed by each ion remains rigid around the ion in the course of lattice vibration. According to this theory,

$$\Delta V = \sum_i U(\mathbf{r} - \mathbf{R}_i - \Delta\mathbf{R}_i) - \sum_i U(\mathbf{r} - \mathbf{R}_i) = \sum_i \Delta\mathbf{R}_i \cdot [-\Delta U_i(\mathbf{r})] \tag{5.8}$$

for the simple case where all ions in the lattice are equivalent. Thus $-\Delta U_i(\mathbf{r})$ stands for $\mathbf{W}_i(\mathbf{r})$ in this case. The deformable-ion theory assumes that the potential for an electron at point **r** of a distorted lattice is the same as that at the point $\mathbf{r} - \Delta\mathbf{R}(\mathbf{r})$ in the undistorted lattice, where $\Delta\mathbf{R}(\mathbf{r})$ is the displacement undergone by point **r** in the distorted lattice. According to this theory,

$$\Delta V = [-\Delta R(\mathbf{r})] \cdot [-\nabla V_0(\mathbf{r})] = \Delta\mathbf{R}(\mathbf{r}) \cdot \nabla V_0(\mathbf{r}). \tag{5.9}$$

Both approximations were used in the development of the theory of metals. We forego details of their applications. Some other treatments of electron–lattice interaction are considered in the following.

a. Deformation Potential Interaction

For acoustic branches of vibration, the lattice deformation may be considered as a strain, provided the wavelength of vibration is long compared to the dimension of a lattice cell. Ions are displaced in the same way as in the presence of a local strain that is nearly uniform over many cells. The effect of a uniform strain on the electron energy

$\epsilon(\mathbf{k})$ may be expressed as

$$\epsilon(\mathbf{k}) = \epsilon_0(\mathbf{k}) + \sum_{i,j} C_{ij}(\mathbf{k}) s_{ij}, \tag{5.10}$$

where $i, j = x, y, z$; s_{ij} is a strain component; and $C_{ij}(\mathbf{k})$ is a "deformation potential constant" that depends on $\mathbf{k}$. The dependence of C_{ij} is easily seen in the following example. For a cubic crystal, the symmetry requires $C_{xx} = C_{yy} = C_{zz}$ and $C_{i,j\neq i} = 0$ for $\mathbf{k} = 0$. On the other hand, for $\mathbf{k} \neq 0$, $C_{i,j\neq i}$ is not required to be zero if $\mathbf{k} \parallel \langle 111 \rangle$, and $C_{yy} = C_{zz} \neq C_{xx}$ is $\mathbf{k} \parallel \langle 100 \rangle$.

Consider the simple case of $\mathbf{k} = 0$ in a cubic crystal. The summation in the above equation reduces to

$$C(s_{xx} + s_{yy} + s_{zz}) = C\Delta, \tag{5.11}$$

where C is the coefficient of uniform strain and Δ is the volume dilatation associated with the strain. Denote by $\delta\mathbf{r}$ the displacement of point $\mathbf{r}$ of the crystal that is produced by the strain. We then have

$$\Delta = \nabla \cdot \delta\mathbf{r}.$$

For a mode $\mathbf{q}p$ of lattice vibration of long wavelength, $\delta\mathbf{r}$ may be identified with the average displacement of the ions in the lattice cell $\mathbf{L} = \mathbf{r}$. Using (2.33) for $\Delta\mathbf{R}_{Lb}$, we get

$$\begin{aligned}(\delta\mathbf{r})_{\mathbf{q}} &= \frac{1}{n}\sum_b (\Delta\mathbf{R}_{Lb})_{\mathbf{q}p} \\ &= \frac{1}{N^{1/2}}\left(\frac{\hbar}{2M_{\mathbf{q}p}\omega_{\mathbf{q}p}}\right)^{1/2} [\mathbf{e}_{\mathbf{q}p} a^{+}_{\mathbf{q}p} \exp(i\mathbf{q}\cdot\mathbf{r}) + e^{*}_{\mathbf{q}p} a_{\mathbf{q}p} \exp(-i\mathbf{q}\cdot\mathbf{r})],\end{aligned} \tag{5.12}$$

where $M_{\mathbf{q}p}$ is defined by

$$\frac{\mathbf{e}_{\mathbf{q}p}}{M^{1/2}_{\mathbf{q}p}} = \frac{1}{n}\sum_b \frac{\mathbf{e}_{\mathbf{q}pb}}{M^{1/2}_b}.$$

The local volume dilatation produced by lattice vibration is then

$$\begin{aligned}\Delta = \nabla \cdot \delta\mathbf{r} = \frac{1}{N^{1/2}} \sum_{\mathbf{q}} \sum_p \left(\frac{\hbar}{2M_{\mathbf{q}p}\omega_{\mathbf{q}p}}\right)^{1/2} i\mathbf{q} \\ \cdot [\mathbf{e}_{\mathbf{q}p} a^{+}_{\mathbf{q}p} \exp(i\mathbf{q}\cdot\mathbf{r}) - \mathbf{e}^{*}_{\mathbf{q}p} a_{\mathbf{q}p} \exp(-i\mathbf{q}\cdot\mathbf{r})].\end{aligned} \tag{5.13}$$

It is seen that only the longitudinal branch, $\mathbf{q} \parallel \mathbf{e} \parallel \mathbf{e}^*$, of the acoustic modes contributes to Δ. Hence the summation over p can be omitted and the subscript p may be dropped. Nonvanishing matrix elements of the interaction $C\Delta$ have the expression

$$\langle \mathbf{k} \mp \mathbf{q} + \mathbf{G}, n_{\mathbf{q}} \pm |C\Delta| \mathbf{k}, n_{\mathbf{q}} \rangle = iN^{-1/2} \left(\frac{\hbar}{2M_{\mathbf{q}}\omega_{\mathbf{q}}} \right)^{1/2} Cq \begin{cases} (n_{\mathbf{q}}+1)^{1/2} \\ n_{\mathbf{q}}^{1/2}. \end{cases} \tag{5.14}$$

The restriction on the change in $\mathbf{k}$ between the initial and final states comes from the intergration of $\exp(i\mathbf{q}\cdot\mathbf{r})$ or $\exp(-i\mathbf{q}\cdot\mathbf{r})$ over $\mathbf{r}$. The deformation potential approximation circumvents the use of $W_{Lb}(\mathbf{r})$, which pertains to the displacement of an individual ion.

It should be noted that electron–lattice interaction is not determined by the local strain if a potential of long range is produced by a lattice deformation. A high density of free carriers, in a metal for example, creates such a situation. The potential variation produced by a variation in local strain leads to a variation in charge density that produces electrostatic potential of long range. The variation in electrostatic potential depends upon the density distribution and the screening effect of the free carriers. Self-consistent treatments show that although longitudinal phonons are mainly responsible for the interaction in cubic crystals, transverse phonons can give rise to U processes, and they may also produce N processes if the Fermi surface of free carriers is not spherical.

b. Interaction with Polar Modes

For a crystal having ions of more than one kind, different electric charges may be attributed to ions of different kinds. Displacements of the ions give rise to a dielectric polarization. The effect is particularly significant for optical modes of long wavelengths such that the ion displacements are nearly the same over many lattice cells. We can then deal with "local" polarization. In a unit cell L, the displacement of positively charged ions as a group against the group of negatively charged ions is

$$\frac{1}{Q}\left(\sum_{\alpha} e_{\alpha}\Delta R_{\alpha L} - \sum_{\beta} |e_{\beta}|\Delta R_{\beta L}\right)_{\mathbf{q}p} = \frac{1}{N^{1/2}}\left(\frac{\hbar}{2M_{\mathbf{q}p}\omega_{\mathbf{q}p}}\right)^{1/2} \times [\mathbf{U}_{\mathbf{q}p}a_{\mathbf{q}p}^{+}\exp(i\mathbf{q}\cdot\mathbf{L}) - \mathbf{U}_{\mathbf{q}p}^{*}a_{\mathbf{q}p}\exp(-i\mathbf{q}\cdot\mathbf{L})], \tag{5.15}$$

where

$$Q = \sum_{\alpha} e_{\alpha} = \sum_{\beta} |e_{\beta}|$$

and

$$\frac{\mathbf{U}_{\mathbf{q}p}}{M_{\mathbf{q}p}^{1/2}} = \frac{1}{Q}\left[\sum_{\alpha} \frac{e_\alpha \mathbf{e}_{\mathbf{q}p,\alpha}}{M_\alpha^{1/2}} - \sum_{\beta} \frac{|e_\beta|\mathbf{e}_{\mathbf{q}p,\beta}}{M_\beta^{1/2}}\right]. \tag{5.16}$$

α and β are indices of positively charged ions and negatively charged ions, respectively, in a lattice cell. The above expression with $\mathbf{L}$ replaced by $\mathbf{r}$ and divided by the volume of the cell will be taken as the local relative displacement.

Longitudinal optical modes, the vibrations with displacement parallel to the wavevector $\mathbf{q}$, produce polarizations $\mathbf{P} \parallel \mathbf{q}$ and electric fields $\mathbf{E} \parallel \mathbf{q}$ with magnetic field $\mathbf{H} = 0$. The electric field associated with the relative displacement of ions may be expressed as

$$\mathbf{E}_{\mathbf{q}p} = -4\pi e_{\mathbf{q}p}^{*}\left(\frac{\mathbf{q}}{q}\right)\frac{1}{\Omega N^{1/2}}\left(\frac{\hbar}{2M_{\mathbf{q}p}\omega_{\mathbf{q}p}}\right)^{1/2} \times \left[d_{\mathbf{q}p}a_{\mathbf{q}p}^{+}\exp(i\mathbf{q}\cdot\mathbf{r}) - d_{\mathbf{q}p}^{*}a_{\mathbf{q}p}\exp(-i\mathbf{q}\cdot\mathbf{r})\right], \tag{5.17}$$

where Ω is the volume of a lattice cell,

$$d_{\mathbf{q}p} = \mathbf{U}_{\mathbf{q}p}\cdot\mathbf{q}/q,$$

and $e_{\mathbf{q}p}^{*}$ is an effective charge of either group of ions for the longitudinal component of the mode $\mathbf{q}p$. $e_{\mathbf{q}p}^{*}$ is given in Chapter VI by (6.149).

An electron at $\mathbf{r}_e$ gives an electric field

$$\frac{\mathbf{D}}{\varepsilon_\infty} = -\frac{e}{|\mathbf{r}-\mathbf{r}_e|^3}(\mathbf{r}-\mathbf{r}_e), \tag{5.18}$$

ε_∞ being the high-frequency dielectric constant. The electrical energy of the lattice vibration and the electron is

$$W = \frac{\varepsilon_\infty}{8\pi}\left[\frac{\mathbf{D}}{\varepsilon_\infty} + \sum_{\mathbf{q}p}\mathbf{E}_{\mathbf{q}p}\right]^2. \tag{5.19}$$

The increase in W given by the electron–lattice interaction is

$$\Delta W = \frac{1}{4\pi}\mathbf{D}\cdot\left(\sum_{\mathbf{q}p}\mathbf{E}_{\mathbf{q}p}\right). \tag{5.20}$$

The interaction potential is then

$$\Delta V = \int \Delta W \, d\mathbf{r} = \sum_{\mathbf{q}p} \frac{1}{4\pi} \int \mathbf{D} \cdot \mathbf{E}_{\mathbf{q}p} \, d\mathbf{r} = \sum_{\mathbf{q}p} \frac{e e^*_{\mathbf{q}p}}{4\pi \Omega N^{1/2}} \left(\frac{\hbar}{2M_{\mathbf{q}p}\omega_{\mathbf{q}p}} \right)^{1/2}$$
$$\times [A_{\mathbf{q}p} a^+_{\mathbf{q}p} \exp(i\mathbf{q}\cdot\mathbf{r}_e) - A^*_{\mathbf{q}p} a_{\mathbf{q}p} \exp(-i\mathbf{q}\cdot\mathbf{r}_e)], \quad (5.21)$$

where

$$A_{\mathbf{q}p} = d_{\mathbf{q}p} \int \frac{\mathbf{r} - \mathbf{r}_e}{(\mathbf{r} - \mathbf{r}_e)^3} \cdot \frac{\mathbf{q}}{q} \exp[i\mathbf{q}\cdot(\mathbf{r} - \mathbf{r}_e)] \, d\mathbf{r}$$
$$= -d_{\mathbf{q}p} \int \frac{\cos\theta}{\rho^2} \exp(iq\rho\cos\theta)\rho^2 \, d\rho \, d\cos\theta \, d\varphi$$
$$= -\frac{4\pi i}{q} d_{\mathbf{q}p}.$$

c. Piezoelectric Interaction with Acoustic Modes

Acoustic modes of lattice vibration can also give rise to dielectric polarization by virtue of the piezoelectric effect. Acoustic waves of long wavelengths are similar to strain waves, and the associated dielectric polarization **P** is determined by the piezoelectric modulus d:

$$\mathbf{P}(\mathbf{r}) = d \cdot S(\mathbf{r}), \quad (5.22)$$

where $S(\mathbf{r})$ is the second-rank tensor of strain. The modulus d is a tensor of third rank. $d = 0$ for a crystal with a center of symmetry and also for crystals of class 432. It is not required to vanish by crystal symmetry for 20 out of 32 classes. For the two cubic classes, $43m$ and 23, the three nonzero components of d are equal, $d_{14} = d_{25} = d_{36}$, and the strain tensor can be obtained from the volume dilatation. The interaction of an electron with the dielectric polarization of the lattice has already been considered in connection with the polar optical modes.

d. Polarons

Electronic states of the rigid lattice do not actually belong to the eigenstates of the crystal, due to the interaction of electrons and lattice vibration. Consider an electronic state $|\mathbf{k}\rangle$ of the rigid lattice. The state $|\mathbf{k}\rangle^{(1)}$ to the first order of perturbation of the interaction H' is

$$|\mathbf{k}\rangle^{(1)} = |\mathbf{k}\rangle + \sum_{\mathbf{q}p} |(\mathbf{k} - \mathbf{q}), 1_{\mathbf{q}p}\rangle \frac{\langle(\mathbf{k} - \mathbf{q}), 1_{\mathbf{q}p}|H'|\mathbf{k}, 0_{\mathbf{q}p}\rangle}{\epsilon_{\mathbf{k}} - \epsilon_{\mathbf{k}-\mathbf{q}} - \hbar\omega_{\mathbf{q}p}}. \quad (5.23)$$

The total number N_p of phonons accompanying the electron is

$$\langle N_p \rangle = \sum_{\mathbf{q}p} {}^{(1)}\langle \mathbf{k}, 0_{\mathbf{q}p} | a^+_{\mathbf{q}p} a_{\mathbf{q}p} | \mathbf{k}, 0_{\mathbf{q}p} \rangle^{(1)}$$

$$= \sum_{\mathbf{q}p} \frac{|\langle (\mathbf{k}-\mathbf{q}), 1_{\mathbf{q}p} | H' | \mathbf{k}, 0_{\mathbf{q}p} \rangle|^2}{\epsilon_{\mathbf{k}} - \epsilon_{\mathbf{k}-\mathbf{q}} - \hbar\omega_{\mathbf{q}p}}. \tag{5.24}$$

The composite particle of an electron and the associated cloud of phonons is sometimes referred to as a dressed electron. If N_p is of the order of unity or larger, the perturbation treatment is not justifiable.

Consider electrons of an effective mass m^*. It can be shown that the deformation potential interaction gives

$$\langle N_p \rangle \sim \frac{1}{2} \frac{m^{*2} C^2}{\hbar^3 \rho c_s} \log\left(\frac{\mathrm{q}_{\max}}{q_c}\right), \tag{5.25}$$

where c_s is the longitudinal sound velocity, ρ is the mass density, and $q_c = 2m^* c_s/\hbar$. $\langle N_p \rangle$ is negligibly small, being of the order of 10^{-12}. On the other hand, electron interaction with polar optical modes of frequency ω_L gives

$$\langle N_p \rangle \sim \frac{e^2}{4\hbar\omega_L} \left(\frac{2m^*\omega_L}{\hbar}\right)^{1/2} \left(\frac{1}{\varepsilon_\infty} - \frac{1}{\varepsilon_0}\right) \equiv \frac{\alpha}{2}, \tag{5.26}$$

which may be quite large for ionic crystals; for example, $\langle N_p \rangle \sim 3$ for the alkali halides. The composite particle of an electron and the associated cloud of polar optical phonons is commonly referred to as a polaron.

The energy of a composite particle, given by the second-order perturbation theory, is

$$\epsilon_{\mathbf{k}} = (\epsilon_{\mathbf{k}})_0 + \sum_{\mathbf{q}p} \frac{|\langle (\mathbf{k}-\mathbf{q}), 1_{\mathbf{q}p} | H' | \mathbf{k}, 0_{\mathbf{q}p} \rangle|^2}{\epsilon_{\mathbf{k}} - \epsilon_{\mathbf{k}-\mathbf{q}} - \hbar\omega_{\mathbf{q}p}}. \tag{5.27}$$

For polarons with $\alpha < 10$,

$$\epsilon_{\mathbf{k}} \sim (\epsilon_{\mathbf{k}})_0 - \alpha\hbar\omega_L - (\alpha/12m^*)\hbar^2 k^2, \tag{5.28a}$$

and

$$m_p^* \equiv \frac{m^*}{1 - \alpha/6} \tag{5.28b}$$

is often called the polaron mass.

2. INTERACTION WITH LATTICE IMPERFECTIONS

There are several types of lattice imperfections that impair perfect periodicity. A grain boundary in a crystal is an imperfection extending over a surface, and a dislocation is an imperfection extending over a line. Imperfections of the following types may be generally referred to as point defects. One, two, or a few neighboring ions missing from the lattice are called a vacancy, a divacancy, and so on. An atom located between normal sites in the lattice is an interstitial atom. A chemical impurity is substitutional if its atoms replace a normal atom of the crystal; the atoms of an impurity can also be interstitial. Electron interaction with the various types of lattice defects has been treated in various connections. The interaction of most common interest is discussed in the following.

Interaction with a charged center. For the interaction with a charge carrier, many point imperfections may be approximated by a charged center. If the screening effect of free carriers is neglected, the interaction potential of such a center is simply the Coulomb potential

$$V = Ze^2/\varepsilon r, \tag{5.29}$$

where Ze is the charge of the center. ε is the dielectric constant of the crystal without the contribution of free carriers, and it may be approximately taken to be the static dielectric function for small $\mathbf{q}$. The differential cross section of the center for an incident plane wave diverges for the polar angle $\theta \to 0$. A lower limit of θ was taken that corresponds to limiting the effect of a center up to a distance equal to one-half the average separation between two neighboring centers. The procedure was based on the assumption that the effect of a center extends only as far as the classical impact parameter is within the volume to be assigned to the center.

Actually, the screening effect of free carriers should be taken into account. Neglecting the dependence of screening on a wavevector, we take

$$V = \frac{Ze^2}{\varepsilon r} \exp\left(\frac{-r}{r_s}\right). \tag{5.30}$$

The parameter r_s, the screening radius, is determined in the following way. A variation of potential produces a variation of density n of the free carriers. Consequently, a space charge $\rho(r)$ is produced.

The Poisson equation of electrostatics is

$$\nabla^2 V = -\frac{4\pi e}{\varepsilon}\rho = -\frac{4\pi e^2}{\varepsilon}(n - n_0), \qquad (5.31)$$

where n_0 is the uniform density of free carriers in the absence of the charged center. If the free carriers have a Maxwell–Boltzmann distribution, we have

$$n = n_0 e^{-V/kT}.$$

The equation becomes

$$\nabla^2 V = \frac{4\pi e^2}{\varepsilon} n_0 (1 - e^{-V/kT}) \sim \frac{4\pi e^2 n_0}{\varepsilon kT} V,$$

the approximation being for the condition $V \ll kT$. If the free carriers are highly degenerate, we have

$$n = \frac{1}{3\pi^2}\frac{2m^*}{\hbar^2}(\epsilon_F - V)^{3/2},$$

where ϵ_F is the Fermi energy. Poisson's equation takes the form

$$\nabla^2 V \sim -\frac{4\pi e^2}{\varepsilon}\left(\frac{dn}{dV}\right)_{V=0} V = \frac{4\pi e^2}{\varepsilon}\left(\frac{1}{3\pi^2}\right)^{2/3}\frac{3m^*}{\hbar^2} n_0^{1/3} V$$

under the condition $V \ll \epsilon_F$. The solution of Poisson's equation gives

$$\left(\frac{1}{r_s}\right)^2 = \frac{4\pi e^2}{\varepsilon}\left(\frac{n_0}{kT}\right) \qquad (5.32a)$$

for free carriers of Maxwell–Boltzmann distribution and

$$\left(\frac{1}{r_s}\right)^2 = \frac{4\pi e^2}{\varepsilon}(3\pi^2)^{-2/3}\frac{3m^* n_0^{1/3}}{\hbar^2} \qquad (5.32b)$$

for degenerate free carriers.

According to (5.30), the screened potential decreases smoothly with distance r. The expression is an approximation. Careful consideration shows that the potential and charge density of conduction electrons exhibit oscillations with r, known as Friedel oscillations, that become significant in the case of a large concentration of conduction electrons.

The ion–ion interaction in a lattice is evidently screened by the conduction electrons. It has been shown by W. Kohn that the effect may produce kinks, infinite $\partial\omega/\partial\mathbf{q}$, in the phonon spectrum at $q = 2k_F$, where $\mathbf{k}_F$ is the wavevector of the Fermi surface of the electrons.

The Kohn anomaly can be useful for the investigation of Fermi surfaces in metals at low temperatures.

3. TEMPERATURE DEPENDENCE OF ENERGY-BAND STATES

A change of temperature affects an electronic energy level ϵ in two ways, through the change in lattice dimension and through the change in lattice vibration. With reference to (5.10), the effect of lattice dilatation can be represented by the deformation potential constants of the particular energy level $\epsilon(m,\mathbf{k})$ for the wavevector $\mathbf{k}$ in the energy band m. Clearly the effect is similar to that of pressure variation. We shall consider here the effect of lattice vibration, which is important and has been the subject of many investigations.

We have treated electron–lattice interaction as a perturbation in Section V.1. Perturbation was considered to be linear in ion displacement. We shall now denote it by H_1. In first-order perturbation theory, the wavefunction of the crystal is perturbed through the matrix elements of H_1 between different unperturbed eigenfunctions, but the energy is not affected. The diagonal matrix elements of the perturbation vanish due to the fact that the expectation value of displacement vanishes for ions in vibration. The second-order perturbation of energy is given by

$$\epsilon'_{m\mathbf{k}} = \sum_{m'\nu\mathbf{q}} \frac{|\langle m',\mathbf{k}\pm\mathbf{q};n_{\nu\mathbf{q}}\mp 1|H_1|m,\mathbf{k};n_{\nu\mathbf{q}}\rangle|^2}{\epsilon_{m\mathbf{k}}-\left(\epsilon_{m',\mathbf{k}\pm\mathbf{q}}\mp\hbar\omega_{\nu\mathbf{q}}\right)}, \tag{5.33}$$

where $n_{\nu\mathbf{q}}$ is the number of phonons of wavevector $\mathbf{q}$ in branch ν. The summation is subject to the requirement that $m',\mathbf{k}\pm\mathbf{q}$ must be unoccupied by an electron. This energy perturbation has come to be known as the effect of self-energy.

Subsequently, the effect of including H_2, electron–lattice interaction quadratic in ion displacements, has been treated. Since the displacement squared has an expectation value in vibration, H_2 produces an energy change ϵ'' in first-order perturbation theory:

$$\epsilon''_{m\mathbf{k}} = \langle m\mathbf{k};n_{\nu\mathbf{q}}|H_2|m\mathbf{k};n_{\nu\mathbf{q}}\rangle. \tag{5.34}$$

An electronic energy level at temperature T is

$$\epsilon_{m\mathbf{k}}(T) = \epsilon^0_{m\mathbf{k}}(T)+\epsilon'_{m\mathbf{k}}(T)+\epsilon''_{m\mathbf{k}}(T). \tag{5.35}$$

The first term refers to the lattice with ions fixed at equilibrium positions; its temperature dependence comes from the effect of lattice dilatation. The second and third terms represent the effect of lattice vibration. ϵ' and ϵ'' are each of the second order of ion displacement. Their temperature dependence comes from the $n_{\nu\mathbf{q}}$'s of thermal equilibrium.

Referring to (5.1), write H_1 and H_2 in the following forms:

$$H_1 = \sum_{Lb} \Delta\mathbf{R}_{Lb} \cdot \mathbf{h}'_{Lb}(\mathbf{r})$$

and

$$H_2 = \sum_{Lb} \Delta\mathbf{R}_{Lb}\Delta\mathbf{R}_{Lb} : \mathbf{h}''_{Lb}(\mathbf{r})\mathbf{h}''_{Lb}(\mathbf{r}).$$

For a translation $\mathbf{t}$ of the ion lattice, $\Delta\mathbf{R}_{Lb} = \mathbf{t}$ is independent of L and b, and the ϵ's are not changed:

$$0 = \Delta\epsilon_{m\mathbf{k}} = \mathbf{tt} : \left[\sum_{Lb} \langle m\mathbf{k}|\mathbf{h}''_{Lb}\mathbf{h}''_{Lb}|m\mathbf{k}\rangle + \sum_{Lb}\sum_{L'b'}\sum_{m'\mathbf{k}'} \frac{\langle m\mathbf{k}|\mathbf{h}'_{Lb}|m'\mathbf{k}'\rangle\langle m'\mathbf{k}'|\mathbf{h}'_{L'b'}|m\mathbf{k}\rangle}{\epsilon_{m\mathbf{k}} - \epsilon_{m'\mathbf{k}'}} \right]. \quad (5.36)$$

The relationship between $h'_{Lb}(\mathbf{r})$ and $h''_{Lb}(\mathbf{r})$ given by equating the content of the square brackets to zero is useful for assessing the relative magnitudes of $\epsilon'_{m\mathbf{k}}$ and $\epsilon''_{m\mathbf{k}}$.

CHAPTER SIX
Properties of Solids

A. GENERAL CONSIDERATIONS

A physical quantity is represented by an operator, an observable. An observable F has an expectation value $\overline{F}$ for a given state of the matter. A quantity of matter consists of a large number of constituents, and the value of a macroscopic quantity of interest is the statistical expectation value $\overline{\overline{F}}$ for an ensemble of systems of the same structure. For systems that have only thermal interaction with the surroundings or for essentially isolated systems, the canonical ensemble is appropriate for the equilibrium situation.

The wavefunction ψ of a system may be expanded in terms of a complete set of orthonormal time-independent functions $\varphi_m(\mathbf{r})$ with time-dependent coefficients $C_m(t)$:

$$\psi = \sum_m C_m \varphi_m. \tag{6.1}$$

The expectation value of an observable F is given by

$$\overline{F} = \sum_{m,n} C_m^* C_n \langle m|F|n\rangle = \sum_{m,n} C_m^* C_n F_{mn}. \tag{6.2}$$

The statistical expectation value of F for an ensemble of M systems is

$$\langle \overline{F} \rangle = \frac{1}{M} \sum_{\alpha=1}^{M} \left[\sum_{m,n} (C_m^* C_n)_\alpha F_{mn} \right] = \sum_{m,n} \left[\frac{1}{M} \sum_{\alpha=1}^{M} (C_m^* C_n)_\alpha \right] F_{mn}. \quad (6.3)$$

Denote

$$\rho_{nm} = \frac{1}{M} \sum_{\alpha=1}^{M} (C_m^* C_n)_\alpha. \quad (6.4)$$

ρ_{nm} is a component of the so-called density matrix ρ for the ensemble. We then have

$$\langle \overline{F} \rangle = \sum_{m,n} \rho_{nm} F_{mn} = \sum_{t} \rho F_{tt} = \mathrm{Tr}(\rho F). \quad (6.5)$$

For an ensemble appropriate for a system in equilibrium,

$$\rho_{nm} = 0 \qquad \text{for} \quad n \neq m, \quad (6.6a)$$

according to the fundamental hypothesis of equal a priori phases, meaning that there is no special selection of phases for the different states of the different members of the ensemble. Hence,

$$\langle \overline{F} \rangle = \sum_{n} \rho_{nn} F_{nn}. \quad (6.6b)$$

This equation is valid whether or not the system is in equilibrium for an operator that has

$$F_{nm} = 0 \qquad \text{for} \quad n \neq 0. \quad (6.6c)$$

According to the Liouville theorem of quantum statistical mechanics, the time dependence of the density matrix is given by

$$\frac{\partial \rho}{\partial t} = \frac{i}{\hbar} (\rho H - H \rho), \quad (6.7a)$$

H being the Hamiltonian operator of the system. The equation can also be written in the form

$$\frac{\partial \rho_{nm}}{\partial t} = \frac{i}{\hbar} \sum_{j} (\rho_{nj} H_{jm} - H_{nj} \rho_{jm}). \quad (6.7b)$$

The time dependence $\partial \rho / \partial t$ is zero for a system in equilibrium.

Consider a system of weakly interacting elements, or "particles." The system reduces to a group of independent particles if the interaction is overlooked. Then the system may be considered an ensemble of one-particle systems, and the subscripts n and m in the previous equations can be replaced by i and i' to denote the one-particle stationary states. This consideration may be applied to the system of electrons, phonons, or other "particles" in a solid. Equation (6.5) leads to

$$\langle \overline{F} \rangle = N \sum_{i,i'} \rho_{ii'} F_{i'i}, \tag{6.8a}$$

where N is the number of particles in the system. For a system in equilibrium and generally for an operator with

$$F_{ii'} = 0 \qquad \text{for} \quad i \neq i', \tag{6.8b}$$

Eq. (6.6) leads to

$$\langle \overline{F} \rangle = N \sum_{i} \rho_{ii} F_{ii} = \sum_{i} f_i F_{ii}. \tag{6.8c}$$

The function of i, f_i, so defined is the distribution function of particles. It gives the probability of finding a particle in state i for a system of N particles. Under equilibrium conditions, f_i is the well-known Fermi–Dirac distribution or Bose–Einstein distribution, depending on the type of particles; $f_i \overline{<} 1$ is required for a group of fermions, whereas f_i may be as large as the total number N of particles for a group of bosons. Consider a transition of the system that decreases the particle occupation of state i with a corresponding increase for state i'. The rate of such transitions is determined by the matrix element $\langle i'|H|i\rangle$ of the one-particle Hamiltonian H. Statistically, the rate is proportional to the availability f_i of particles in state i and the possibility of increasing particles in state i'. The latter is unity for bosons and $1 - f_{i'}$ for fermions.

The properties of a substance may be classified according to whether or not they pertain to equilibrium states of the substance. The specific heat, the thermal expansion coefficient, and the elastic constants are examples of "equilibrium properties." For substances that can be in equilibrium under an applied constant electric or magnetic field, the dielectric or magnetic susceptibility is also an equilibrium property. For properties of this category, Eq. (6.6b) is valid, and (6.8a) applies for a substance consisting of weakly interacting particles. Properties of the second category are not defined in terms of equilibrium conditions of the substance. Some examples of this kind of property are electrical conductivity and thermal conductivity, which characterize, respectively, the flow of charge and the flow of heat. To treat properties of this category, we should begin with the density matrix for equilibrium at $t = 0$, $\rho_{nm}(0) = 0$

for $n \neq m$, use Eq. (6.7b) to calculate $\partial \rho_{nm}/\partial t$ due to the external perturbation H_{ext} applied at $t = 0$, and calculate the statistical expectation value of the pertinent observable according to (6.5) with $\rho_{nm}(t)$. The treatment is simplified if the one-particle approximation of the preceding paragraph is adopted. An additional approximation is involved if (6.8c) is used for an operator not characterized by (6.8b).

1. BOLTZMANN'S TRANSPORT EQUATION

This equation originated in the kinetic theory of gases. It may be applied to a system of electrons, excitons, phonons, polarons, and so on. A "local" distribution $f(\mathbf{k}, \mathbf{r}, t)$ is considered, which refers to a small region dr^3 of sufficient extent for defining the eigenstates $\mathbf{k}$ of a particle. The Boltzmann equation states

$$\frac{\partial f}{\partial t} = \left(\frac{\partial f}{\partial t}\right)_{\text{drift}} + \left(\frac{\partial f}{\partial t}\right)_{\text{collision}}, \tag{6.9}$$

where $(\partial f/\partial t)_{\text{drift}}$ is the rate of change due to the drift motion of the particles, and $(\partial f/\partial t)_{\text{collision}}$ is the rate of change due to the various interactions experienced by a particle. In the left-hand side of the equation, $\partial f/\partial t = 0$ for a system in a constant, steady state, and $\partial f/\partial t \propto e^{i\omega t}$ for a system under a steady periodic perturbation.

The one-electron eigenstate is characterized by the wavevector $\mathbf{k}$. The "local" distribution function f stands for electron distribution among the states $|k\rangle$ at a location $\mathbf{r}$. We have for the "drift"

$$\left(\frac{\partial f}{\partial t}\right)_d = -\left[\frac{d\mathbf{k}}{dt} \cdot \nabla_{\mathbf{k}} f + \mathbf{v} \cdot \nabla_{\mathbf{r}} f\right]. \tag{6.10}$$

With the application of a constant electric field $\mathbf{E}$ and a constant magnetic field $\mathbf{B}$, we have, according to (3.34),

$$\frac{d\mathbf{k}}{dt} = \frac{e}{\hbar}\mathbf{E} + \frac{1}{c}\mathbf{v} \times \mathbf{B}. \tag{6.11}$$

The distribution may be considered a sum:

$$f = f_0 + \Delta f, \tag{6.12}$$

where f_0 corresponds to local equilibrium. Δf is usually much smaller than f_0. Now,

$$\nabla_{\mathbf{k}} f_0 = \frac{\partial f_0}{\partial \epsilon} \nabla_{\mathbf{k}} \epsilon = \frac{\partial f_0}{\partial \epsilon} \hbar \mathbf{y}. \tag{6.13}$$

For the effect of the applied electric field $\mathbf{E}$ on $(\partial f/\partial t)_{\mathrm{d}}$, it is sufficient to take $\nabla_{\mathbf{k}} f_0$ for $\nabla_{\mathbf{k}} f$. For an applied magnetic field, however, $\nabla_{\mathbf{k}} f_0$ gives no effect on $(\partial f/\partial t)_d$, since $(\mathbf{v}\times\mathbf{B})\cdot\mathbf{v}=0$. Therefore,

$$\frac{d\mathbf{k}}{dt}\cdot\nabla_{\mathbf{k}} f=\frac{e}{\hbar}\mathbf{E}\cdot\nabla_{\mathbf{k}} f_0+\frac{e}{\hbar c}(\mathbf{v}\times\mathbf{B})\cdot\nabla_{\mathbf{k}}(\Delta f). \tag{6.14}$$

Consider now the second term on the right-hand side of (6.10). Using f_0 for f, we get

$$\nabla_{\mathbf{r}} f\sim\nabla_{\mathbf{r}} f_0=\frac{df_0}{d(\epsilon-\epsilon_{\mathrm{F}})/kT}\nabla_{\mathbf{r}}\left(\frac{\epsilon-\epsilon_{\mathrm{F}}}{kT}\right)=T\frac{\partial f_0}{\partial\epsilon}\nabla_{\mathbf{r}}\left(\frac{\epsilon-\epsilon_{\mathrm{F}}}{T}\right). \tag{6.15}$$

Depending on the phenomena of interest, either one or both of $\epsilon-\epsilon_{\mathrm{F}}$ and T may vary with $\mathbf{r}$. In summary, we have

$$\begin{aligned}\left(\frac{\partial f}{\partial t}\right)_{\mathrm{d}}&=-e\frac{\partial f_0}{\partial\epsilon}\mathbf{v}\cdot\mathbf{E}-\frac{e}{\hbar c}(\mathbf{v}\times\mathbf{B})\cdot\nabla_{\mathbf{k}}(\Delta f)\\&\quad-T\frac{\partial f_0}{\partial\epsilon}\mathbf{v}\cdot\nabla_{\mathbf{r}}\left(\frac{\epsilon-\epsilon_{\mathrm{F}}}{T}\right).\end{aligned} \tag{6.16}$$

Consider now the term $(\partial f/\partial t)_{\mathrm{c}}$ of "collision." The Boltzmann equation was originally developed for the classical kinetic theory of gases. A gas molecule of momentum $\mathbf{p}$ changes its $\mathbf{p}$ upon collision with another molecule. Also, an electron in a one-electron state makes transitions to other states, due to its interaction with other constituents of the substance. Let V_{I} be a term of the Hamiltonian that represents an interaction. According to the perturbation theory, the interaction produces transitions from state $|\mathbf{k}\rangle$ to state $|\mathbf{k}'\rangle$ with a probability per unit time given by

$$\frac{\partial|C_{\mathbf{k}',\mathbf{k}}|^2}{\partial t}=\frac{2}{\hbar}|\langle\mathbf{k}'|V_{\mathrm{I}}|\mathbf{k}\rangle|^2\frac{\sin(xt)}{x}, \tag{6.17}$$

where

$$x=(E'-E)/\hbar,$$

E and E' being the energy of the substance before and after the transition, respectively. A transition is possible provided there are electrons in the initial state $|\mathbf{k}\rangle$ and vacancies in the final state $|\mathbf{k}'\rangle$. Therefore, we have

$$\left(\frac{\partial f_{\mathbf{k}}}{\partial t}\right)_{\mathrm{c}}=\int\frac{d\mathbf{k}'}{(2\pi)^3}\left[\frac{\partial|C_{\mathbf{k},\mathbf{k}'}|^2}{\partial t}f_{\mathbf{k}'}(1-f_{\mathbf{k}})-\frac{\partial|C_{\mathbf{k}',\mathbf{k}}|^2}{\partial t}f_{\mathbf{k}}(1-f_{\mathbf{k}'})\right]. \tag{6.18}$$

In view of this expression of $(\partial f/\partial t)_c$, the Boltzmann equation is an integro-differential equation instead of a differential equation.

2. RELAXATION TIME

With the collision term $(\partial f/\partial t)_c$ in the form of an integral, the Boltzmann equation is very complicated and difficult to apply. In practice, the relaxation time approximation is made:

$$\left(\frac{\partial f_{\mathbf{k}}}{\partial t}\right)_c = -\frac{f_{\mathbf{k}} - (f_{\mathbf{k}})_0}{\tau(\mathbf{k})}, \tag{6.19}$$

where $\tau(\mathbf{k})$ is the relaxation time of the distribution function for state $|\mathbf{k}\rangle$. In this approximation, the Boltzmann equation reduces to

$$e\frac{\partial f_0}{\partial \epsilon}\left[\mathbf{v}\cdot\mathbf{E} + T\mathbf{v}\cdot\nabla_{\mathbf{r}}\left(\frac{\epsilon-\epsilon_F}{T}\right)\right] + \frac{e}{\hbar c}(\mathbf{v}\times\mathbf{B})\cdot\nabla_{\mathbf{k}}(\Delta f) + \frac{\Delta f}{\tau} = 0, \tag{6.20}$$

which is a differential equation for Δf. It becomes an algebraic equation if there is no applied magnetic field.

Actually, $(\partial f_{\mathbf{k}}/\partial t)_c$ involves $f_{\mathbf{k}'}$ of all states $|\mathbf{k}'\rangle$, as shown by (6.18). Therefore approximation (6.19) requires justification, and the kind of phenomena for which a given $\tau(\mathbf{k})$ applies should be specified. We present in the following the expressions of $\tau(\mathbf{k})$ derived for some important mechanisms of interaction, for the electrical and thermal conduction phenomena produced by an applied electric field and/or temperature gradient. The charge carriers, conduction electrons or holes, in one energy band or in one valley of a multivalley energy band are considered. In the absence of interband or intervalley transitions, the conductions of various bands or valleys are simply additive. The effective mass approximation, $\epsilon = \hbar^2 k^2/2m^*$ for the carriers is assumed.

Interaction with acoustic phonons. Of the acoustic vibrations of the lattice, the longitudinal modes are primarily responsible for the scattering of electrons. Take the maximum longitudinal phonon given by (2.53):

$$(6\pi^2 N)^{1/3}\hbar c_l = k\Theta_l,$$

where the sound velocity c_l and the Debye temperature Θ_l in this case refer exclusively to longitudinal vibrations. It can be shown that for nearly degenerate carriers τ is given by

$$\frac{1}{\tau} = \frac{9\pi^3}{8\sqrt{2}} \frac{\hbar^2 N}{(m^*)^{1/2} M k \Theta_l} C^2 \frac{T}{\Theta_l} \epsilon^{-3/2} \quad \text{for} \quad T > \Theta \tag{6.21}$$

where M is the ionic mass of the assumed monatomic crystal and C is the deformation potential constant in approximation (5.11). For nondegenerate carriers, we have

$$\frac{1}{\tau} = \frac{m^{*2}}{\hbar^4 c_l^2 \rho} C^2 kT \frac{\hbar k}{m^*}, \tag{6.22}$$

where ρ is the mass density of the crystal and $\hbar k/m^* = v$ is the speed of the carrier.

Interaction with polar optical phonons. The expression derived is

$$\frac{1}{\tau} = \frac{8\pi e^2 m^*}{3\hbar} (\hbar\omega_l)^{1/2} \frac{1}{4\pi} \left(\frac{1}{\varepsilon_\infty} - \frac{1}{\varepsilon_0} \right) \exp\left(\frac{-\hbar\omega_l}{kT} \right) \quad \text{for} \quad kT \ll \hbar\omega_l, \tag{6.23}$$

$\hbar\omega_l$ being the energy of an optical phonon.

Interaction with piezoelectric acoustic vibrations. In a crystal with the zinc blende structure, the scattering is represented by

$$\frac{1}{\tau} = \frac{8\pi e^2 m^{*1/2}}{\sqrt{2}\hbar^2 \varepsilon_0^2} d_{14}^2 \left[\frac{16}{13(C_{11} + 2C_{12} + 4C_{44} + 16\pi d_{14}^2/\varepsilon_0)} + \frac{6}{13(C_{44} + 4\pi d_{14}^2/\varepsilon_0)} \right] \frac{kT}{\epsilon^{1/2}}, \tag{6.24}$$

where C_{ij} denotes an elastic modulus and d_{14} is the piezoelectric modulus. In comparison with the scattering by polar optical modes, the effect of this mechanism may become important at low temperatures.

Interaction with a charged center Ze. The relaxation time derived for this mechanism is

$$\frac{1}{\tau} = \frac{Z^2 e^4}{\varepsilon^2 m^{*2} v^3} 2\pi \left[\ln\left(1 + \frac{1}{y}\right) - \frac{1}{1+y} \right], \tag{6.25}$$

where

$$y = \frac{\hbar^2 (1/r_s)^2}{2m^*} (2m^* v^2)^{-1}$$

and v is the speed of the carrier. The screening radius r_s is given by (5.40). Usually, the crystal of interest contains more than one charged center. Provided the concentration of charged centers is small enough to avoid the interference of scatterings by different centers, the $1/\tau$ for a group of centers is simply the sum of that of individual, isolated centers.

In addition to the interactions considered above, many other scattering mechanisms, carrier–carrier scattering and scatterings by particular kinds of crystal imperfections, may be significant. We shall not present here the approximate expressions of $1/\tau$ obtained for some of the additional mechanisms. When several kinds of interactions are to be taken into account, it is usually assumed that

$$\frac{1}{\tau_{\mathbf{k}}} = \sum_i \left(\frac{1}{\tau_{\mathbf{k}}}\right)_i, \tag{6.26}$$

where i is an index for individual types of scattering. This is an approximation insofar as each $(1/\tau_{\mathbf{k}})_i$ depends upon f, which is determined by the combined effect of various types of scattering.

B. THERMAL AND ELECTRICAL PROPERTIES

1. ELECTRONIC SPECIFIC HEAT

Specific heat concerns equilibrium conditions of a substance. The specific heat at constant volume is given by

$$C_v = (\partial E/\partial T)_v, \tag{6.27}$$

where E is the internal energy of the substance at equilibrium. According to (6.8c), the electronic contribution to the energy is

$$E = \sum_i \epsilon_i f_i = \int \epsilon f(\epsilon)\, d\epsilon.$$

The one-electron stationary states i being quasi-continuous, the summation may be replaced by integration. We shall consider C_v per unit volume of the substance, in which case the summation and the integration are over states in a unit volume. $f(\epsilon)$ is the Fermi–Dirac distribution function:

$$f(\epsilon) = \frac{2}{[\exp(\epsilon - \epsilon_F)/kT] + 1} \tag{6.28}$$

for electrons of both spins. The states with $\epsilon \ll (\epsilon_F - kT)$ contribute a constant value to E and therefore are not significant for C_v.

For a metal, the energy bands below the conduction band have $\epsilon \ll (\epsilon_F - kT)$ for temperatures of interest. Only the conduction electrons with a concentration $n = \sum_i f_i$ practically independent of temperature need be considered. Consider the simple case of effective-mass approximation for the conduction band. Take the minimum energy of the band to be zero. By neglecting terms of the first and higher orders in kT/ϵ_F, we get

$$C_v \approx \tfrac{1}{2}\pi^2 nk[kT/\epsilon_F(0)], \tag{6.29}$$

where $\epsilon_F(0)$ is the Fermi energy at $T = 0$. It is interesting to note the difference between this result and $C_v = \frac{3}{2}nk$ given by the classical Maxwell–Boltzmann distribution function.

For a semiconductor, both conduction electrons and holes should be taken into account. Both types of carriers usually have small concentrations. However, the variation in concentration with temperature should be considered in determining C_v.

2. THERMAL AND ELECTRICAL CONDUCTIVITIES

These and other transport properties concern the flow of charge and the flow of energy in the material. The material is not in an equilibrium condition, and the Boltzmann equation can be used to treat the relevant phenomena. The lattice thermal conductivity has been discussed in Section II.7, and electrical conduction of the lattice involves the drift of ions. We will be concerned here with properties of the electronic system. The electric current density $\mathbf{j}$ and the heat current density $\mathbf{w}$ result from the departure Δf from the equilibrium distribution function f_0:

$$\begin{aligned} \mathbf{j} &= e\int \mathbf{v} f \frac{d\mathbf{k}}{4\pi^3} = e\int \mathbf{v}\Delta f \frac{d\mathbf{k}}{4\pi^3}, \\ \mathbf{w} &= \int \epsilon \mathbf{v} f \frac{d\mathbf{k}}{4\pi^3} = \int \epsilon \mathbf{v}\Delta f \frac{d\mathbf{k}}{4\pi^3}. \end{aligned} \tag{6.30}$$

The current $\mathbf{j}/e$ is often characterized by the mobility of the carriers;

$$\mu = |\mathbf{j}/en|, \tag{6.31}$$

n being the carrier concentration.

Consider that there is no applied magnetic field. Using Δf given by (6.20), we get the following expressions for the current densities:

$$\begin{aligned} \mathbf{j} &= eK_1\left(e\mathbf{E} - T\nabla_\mathbf{r}\frac{\epsilon_F}{T}\right) - eK_2\left(\frac{1}{T}\nabla_\mathbf{r}T\right), \\ \mathbf{w} &= K_2\left(e\mathbf{E} - T\nabla_\mathbf{r}\frac{\epsilon_F}{T}\right) - K_3\left(\frac{1}{T}\nabla_\mathbf{r}T\right). \end{aligned} \tag{6.32}$$

The K's are tensors with components

$$(K_n)_{ij} = -\frac{\partial f_0}{\partial \epsilon}\epsilon^{(n-1)}v_iv_j\tau\frac{d\mathbf{k}}{4\pi^3}. \tag{6.33}$$

For a crystal at a uniform temperature, the electrical conductivity tensor is given by

$$\sigma = e^2K_1. \tag{6.34}$$

The thermal conductivity tensor, under the condition $\mathbf{j} = 0$, is given by

$$\kappa = -\frac{\mathbf{w}}{\nabla_\mathbf{r}T} = \frac{K_1K_3 - K_2^2}{K_1T}. \tag{6.35}$$

Consider the following simplification: a spherical energy band with a constant effective mass and an isotropic relaxation time $\tau(\epsilon)$. The K's and consequently the conductivities are then unit tensors. For charge carriers having the Maxwell–Boltzmann distribution, a frequent situation in semiconductors, we get

$$\sigma = e^2n\langle v^2\tau\rangle/3kT \qquad \text{or} \qquad \mu = |e|\langle v^2\tau\rangle/3kT. \tag{6.36}$$

For degenerate carriers, as in metals and sometimes in semiconductors, the electrical conductivity is given by

$$\sigma = \tfrac{1}{3}e^2[\langle v^2\tau\rangle\rho]_{\epsilon_F} + 0[(kT/\epsilon_F)^2] = e^2n\tau_F/m^* \tag{6.37}$$

where ρ is the density of states at the Fermi surface and $\langle\ \rangle$ denotes averaging over the Fermi surface. The thermal conductivity is zero for perfectly degenerate carriers. Taking κ in the lowest (second) order of kT/ϵ_F, we get the Wiedemann–Franz ratio

$$L \equiv \frac{\kappa}{\sigma T} = \frac{\pi^2}{3}\left(\frac{k}{e}\right)^2 \tag{6.38}$$

to be equal to a universal constant under the approximations made.

3. THERMOELECTRIC POWER

The thermoelectric effects, the Thomson effect of a substance and the Peltier effect and thermoelectric power for two substances, are thermodynamically related. Consider thermoelectric power. Two substances A and B form a loop, with the two junctions at different temperatures T' and $T'' > T'$, respectively. The loop is open, with the two terminals in A and at the same temperature T_0. Let subscripts 1 and 2 refer to the terminals connected with the cold and hot junctions, respectively. The thermoelectric power Q_{AB} of the thermocouple is

$$Q_{\mathrm{AB}} = \frac{d(V_2 - V_1)}{d(T'' - T')} = \frac{d}{dT} \int_{\mathbf{r}_2}^{\mathbf{r}_1} \mathbf{E} \cdot d\mathbf{r}. \tag{6.39}$$

The temperature variation gives rise to an electric field $\mathbf{E}$ in each substance. According to (6.32), for our case of $\mathbf{j} = 0$, we get

$$e\mathbf{E} = \nabla_{\mathbf{r}} \epsilon_{\mathrm{F}} - \frac{\epsilon_{\mathrm{F}}}{T} \nabla_{\mathbf{r}} T + \frac{K_2}{K_1 T} \nabla_{\mathbf{r}} T = \nabla_{\mathbf{r}} \epsilon_{\mathrm{F}} + \frac{K_2 - \epsilon_{\mathrm{F}} K_1}{K_1 T} \nabla_{\mathbf{r}} T.$$

The first term on the right-hand side gives zero for the integration between the two terminals, which are in the same substance and at the same temperature. Therefore, we get

$$\begin{aligned}
\frac{Q_{\mathrm{AB}}}{e} &= \int_{\mathbf{r}_1}^{\mathbf{r}_2} \frac{K_2 - \epsilon_{\mathrm{F}} K_1}{K_1 T} \nabla_{\mathbf{r}} T \cdot d\mathbf{r} \\
&= \int_{T_0}^{T''} \left(\frac{Q}{e}\right)_{\mathrm{A}} dT + \int_{T''}^{T'} \left(\frac{Q}{e}\right)_{\mathrm{B}} dT + \int_{T'}^{T_0} \left(\frac{Q}{e}\right)_{\mathrm{A}} dT \\
&= \int_{T'}^{T''} \left[\left(\frac{Q}{e}\right)_{\mathrm{A}} - \left(\frac{Q}{e}\right)_{\mathrm{B}}\right] dT.
\end{aligned} \tag{6.40}$$

The quantity

$$Q = \frac{1}{e} \frac{K_2 - \epsilon_{\mathrm{F}} K_1}{K_1 T} \tag{6.41}$$

is called the absolute thermoelectric power of the substance. The sign of Q is often used to determine the sign of charge e of charge carriers in a semiconductor, $(K_2 - \epsilon_{\mathrm{F}} K_1)/K_1$ being normally positive. Furthermore, the magnitude of Q is sometimes used to estimate ϵ_{F} and the carrier concentration.

4. ISOTHERMAL, GALVANOMAGNETIC PROPERTIES

Introduce the notation

$$\Phi = \frac{\Delta f}{\partial f_0/\partial \epsilon}, \tag{6.42}$$

in terms of which the density of electric current is given by

$$\mathbf{j} = \frac{e}{\hbar}\int \nabla_{\mathbf{k}}\epsilon\,\Phi \frac{\partial f_0}{\partial \epsilon}\frac{d\mathbf{k}}{4\pi^3}. \tag{6.43}$$

Under isothermal conditions, we get from (6.20)

$$\frac{e}{\hbar}\mathbf{E}\cdot\nabla_{\mathbf{k}}\epsilon - \frac{e}{\hbar^2 c}\mathbf{B}\cdot\Omega\Phi + \frac{1}{\tau}\Phi = 0, \tag{6.44}$$

where the operator Ω is

$$\Omega = \nabla_{\mathbf{k}}\epsilon \times \nabla_{\mathbf{k}}.$$

The solution of this equation in ascending powers of $\mathbf{B}$ is

$$\Phi = -\frac{e}{\hbar}\left[\tau\mathbf{E}\cdot\nabla_{\mathbf{k}}\epsilon + \frac{e}{\hbar^2 c}\tau\mathbf{B}\cdot\Omega(\tau\mathbf{E}\cdot\nabla_{\mathbf{k}}\epsilon) + \left(\frac{e}{\hbar^2 c}\right)^2 \tau\mathbf{B}\cdot\Omega\{\tau\mathbf{B}\cdot\Omega(\tau\mathbf{E}\cdot\nabla_{\mathbf{k}}\epsilon)\} + \cdots\right]. \tag{6.45}$$

The solution is useful for sufficiently weak magnetic fields such that the terms are progressively smaller with increasing order of $\tau\mathbf{B}$.

For the simple case of $\epsilon = \hbar^2 k^2/2m^*$ and $\tau(\epsilon)$ dependent only on ϵ, we have, to the second order of B,

$$\mathbf{j} = \sigma_B\mathbf{E} - \omega_c(\tau\sigma)_B\hat{B}\times\mathbf{E} + \omega_c^2(\tau^2\sigma)_B(\hat{B}\cdot\mathbf{E})\hat{B}, \tag{6.46}$$

where $\omega_c = eB/m^*c$, σ is the electrical conductivity for $B = 0$, and $(\tau^m\sigma)_B$ stands for multiplying the integrand of σ by $\tau^m/[1+(\omega_c\tau)^2]$.

It should be emphasized that Eq. (6.44) applies under the condition that the electronic states have Bloch functions characterized by wavevectors $\mathbf{k}$. As shown in Section IV.A.3, the electronic states under an applied magnetic field are different. However, if the time τ between collisions suffered by the electron is short in comparison with the cyclotron period $1/\omega_c = (eB/M^*c)^{-1}$, for example, $\tau\omega_c \ll 1$, Eq. (6.44) may be a fair approximation. In fact, under such circumstances, the terms with increasing order of $\tau\mathbf{B}$ in expression (6.45) are indeed progressively smaller.

In the treatment of a galvanomagnetic property, it is sometimes more convenient to deal with the resistivity tensor ρ instead of the conductivity tensor σ. The components of these tensors have the well-known relationship

$$\rho_{ij} = (\sigma^{-1})_{ij} = (-1)^{i+j}\Delta_{ji}/D, \tag{6.47}$$

where D is the determinant of the tensor σ and Δ_{ji} is the determinant that remains after deleting the jth row and the ith column of σ.

a. Hall Effect

With a current density $\mathbf{j} \parallel \hat{x}$ under an applied magnetic field $\mathbf{B} \parallel \hat{z}$, there is a component E_y of the electric field. The Hall effect concerns the difference $E_y(B) - E_y(0)$. The quantity

$$R = \frac{E_y(B) - E_y(0)}{jB} = \frac{\rho_{yx}(B) - \rho_{yx}(0)}{B} \tag{6.48}$$

is called the Hall coefficient. The component of resistivity ρ_{xy} is related to the conductivity tensor according to (6.47), and the conductivity is given by $\mathbf{j}$ in terms of $\mathbf{E}$. The Hall coefficient can therefore be obtained by calculating $\mathbf{j}$ with $\mathbf{B}$ and with $\mathbf{B} = 0$ in expression (6.45) for Φ.

Consider the simple case of $\epsilon = \hbar^2k^2/2m^*$ and $\tau(\epsilon)$, for which expression (6.46) for $\mathbf{j}$ applies. In this case, $E_y(0) = 0$ and

$$R = \frac{E_y(B_z)}{j_xB_z} = \frac{e}{m^*c}(\tau\sigma)_B\{\sigma_B^2 + [\omega_c(\tau\sigma)_B]^2\}^{-1}. \tag{6.49}$$

Under sufficiently weak magnetic fields,

$$R \sim \frac{e}{m^*c}\frac{(\tau\sigma)_B}{\sigma_B^2} \sim \frac{e}{m^*c}\frac{\tau\sigma}{\sigma^2}. \tag{6.50}$$

For perfectly degenerate charge carriers, we get

$$R = \frac{e}{m^*c}\frac{\tau_F}{\sigma} = \frac{e}{m^*c}\tau_F\frac{m^*}{ne^2\tau_F} = \frac{1}{ecn}. \tag{6.51}$$

On the other hand, charge carriers with a Maxwell–Boltzmann distribution have

$$R = \frac{e}{m^*c}\langle v^2\tau^2\rangle\left[\frac{ne}{3kT}\langle v^2\tau\rangle^2\right]^{-1} = \frac{\mu_B/\mu}{ecn}, \tag{6.52}$$

where

$$\mu_B \equiv |R|\sigma c = \frac{e}{m^*}\frac{\langle v^2\tau^2\rangle}{v^2\tau}$$

is the so-called Hall mobility.

The magnitude of the Hall coefficient serves to determine the carrier concentration n. For nondegenerate charge carriers, the ratio μ_B/μ depends on the scattering mechanisms; for example, $\mu_B/\mu = 3\pi/8$ is the scattering by acoustic modes of lattice vibration is dominant, and $\mu_B/\mu = 1.79$ is the scattering by charged centers is dominant. In addition to its magnitude, the sign of the Hall coefficient is also important; it gives the sign of the charge of the carriers.

b. Magnetoresistance

The change of resistivity tensor brought about by an applied magnetic field may be referred to as the magnetoresistance effect. It has become a practice to call the ratio

$$\frac{[E_{(\| \jmath)}/\jmath]_{\mathbf{j},\mathbf{B}} - [E_{(\| \jmath)}/\jmath]_{\mathbf{j},0}}{[E_{(\| \jmath)}/\jmath]_{\mathbf{j},0}} \tag{6.53}$$

the magnoresistance—longitudinal magnetoresistance for $\mathbf{j} \parallel \mathbf{B}$ and the transverse magnetoresistance for $\mathbf{j} \perp \mathbf{B}$.

Consider the simple case of $\epsilon = \hbar^2k^2/2m^*$ and $\tau(\epsilon)$. For $\mathbf{j} \parallel \mathbf{B}$, Eq. (6.46) becomes

$$\mathbf{j} = \sigma_B\mathbf{E} + \omega_c^2(\tau^2\sigma)(\hat{B}\cdot\mathbf{E})\hat{B}$$

due to the fact that $\mathbf{E} \parallel \mathbf{B}$; a component of $\mathbf{E}$ that is perpendicular to $\mathbf{B}$ would give a component of $\mathbf{j}$ that is perpendicular to $\mathbf{B}$. Hence,

$$\jmath = [\sigma_B + \omega_c^2(\tau^2\sigma)_B]E \equiv \sigma_l E. \tag{6.54}$$

The longitudinal magnetoresistance is given by

$$\frac{1/\sigma_l - 1/\sigma}{1/\sigma} = \frac{\sigma}{\sigma_B + \omega_c^2(\tau^2\sigma)_B} - 1. \tag{6.55}$$

The quantity is zero for a perfectly degenerate distribution of charge carriers. It vanishes in general if τ is a constant independent of ϵ.

Consider now the case of $\mathbf{j} \perp \mathbf{B}$. Since

$$\mathbf{j}\cdot\hat{B} = \sigma_B\mathbf{E}\cdot\hat{B} + \omega_c(\tau^2\sigma)_B\mathbf{E}\cdot\hat{B} = 0,$$

we have $\mathbf{E} \perp \mathbf{B}$, and consequently

$$\jmath = \sigma_B\mathbf{E}\cdot\hat{\jmath} - \omega_c(\tau\sigma)_B(\hat{B}\times\mathbf{E})\cdot\hat{\jmath}. \tag{6.56}$$

Now

$$\hat{\jmath} \cdot (\hat{B} \times \mathbf{E}) = \mathbf{E} \cdot (\hat{\jmath} \times \hat{B}) = -RjB.$$

Using expression (6.49) for the Hall coefficient R, we get

$$j \cdot (\hat{B} \times \mathbf{E}) = -\omega_c(\tau\sigma)_B\{\sigma_B^2 + [\omega_c(\tau\sigma)_B]^2\}^{-1} j. \tag{6.57}$$

It follows that

$$\begin{aligned} j &= \sigma_B \mathbf{E} \cdot \hat{\jmath} + [\omega_c(\tau\sigma)_B]^2\{\sigma_B^2 + [\omega_c(\tau\sigma)_B]^2\}^{-1} j \\ &= \sigma_B\{1 + [\omega_c(\tau\sigma)_B]^2/\sigma_B^2\}\mathbf{E} \cdot \hat{\jmath} \equiv \sigma_t \mathbf{E} \cdot \hat{\jmath}. \end{aligned} \tag{6.58}$$

The transverse magnetoresistance

$$\frac{1/\sigma_t - 1/\sigma}{1/\sigma} = \frac{\sigma}{\sigma_B\{1 + [\omega_c(\tau\sigma)_B]^2/\sigma_B^2\}} - 1 \tag{6.59}$$

is zero for perfectly degenerate carriers, and it is in general zero if the relaxation time is a constant.

Electron Trajectory. The electron trajectory concept has been useful for the investigation of a more general $\epsilon(\mathbf{k})$, particularly for the study of the Fermi surface in a metal, which is not a sphere in the $\mathbf{k}$ space. The applied magnetic field $\mathbf{B}$ should be sufficiently weak that the electron states, $\epsilon(\mathbf{k})$, are not significantly altered and $d\mathbf{k}/dt = (e/\hbar c)\mathbf{v} \times \mathbf{B}$ for a wave packet is meaningful. Under such conditions, a wave packet describes a trajectory, an orbit, in a plane normal to the applied $\mathbf{B}$. The orbit in the $\mathbf{k}$ space is on a surface of constant energy ϵ, since the velocity, $\mathbf{v} = \nabla_{\mathbf{k}}\epsilon/\hbar$, is normal to the surface.

In general, the Fermi surface may be intersected by the boundary of the reduced Brillouin zone. Electron orbits on the surface can be divided into three categories: closed, open, and extended. There are two types of closed orbits; orbits of one type are closed within the reduced zone, whereas those of the other type pass through the first zone and some of its immediate neighbors in the extended zone scheme. An orbit of the first type encloses states of lower energy, and it may be classified as an electron orbit; an orbit of the second type encloses states of higher energy, and it may be classified as a hole orbit. An electron in orbits of the two types rotate in opposite directions around the magnetic field. As to orbits of the other two types, an open orbit describes a path that is wavy but generally straight in the structure of an extended zone, and an extended orbit does not have a generally straight direction and cannot be contained in one zone, no matter how the zone is placed.

The calculation of the conductivity tensor for degenerate carriers is very complicated if the Fermi surface is not simply spherical. The characteristics of different categories of orbits are helpful to consider. Usually, a majority of carriers close to the Fermi surface have closed orbits. It can be shown that the conductivity contributed by such carriers has the form

$$\sigma_{ij} = \begin{vmatrix} A/B^2 & B/B & C/B \\ -B/B & D/B & E/B \\ -C/B & -E/B & F \end{vmatrix} \tag{6.60}$$

to the lowest order in $1/B$, the direction of $\mathbf{B}$ being taken as $\hat{z}$. On the other hand, the contribution of carriers having open orbits tend to the form

$$\sigma_{ij} \xrightarrow[B\to\infty]{} \begin{vmatrix} K\sin^2\theta & K\sin\theta & 0 \\ K\sin\theta\cos\theta & K\cos^2\theta & 0 \\ 0 & 0 & 0 \end{vmatrix}, \tag{6.61}$$

where θ is the angle between the electric field $\mathbf{E}$ and the orbit axis.

Schubnikov–de Haas Effect. An interesting phenomenon is the oscillation of magnetoresistance with the variation of B in the high field range. The effect was first observed in bismuth at low temperatures by Schubnikov and de Haas. Evidently, the effect is a consequence of the quantization under magnetic field, which is discussed in Section IV.A.3 only for the effective-mass approximation. A quantitative treatment of the effect would be quite complicated.

C. ELECTROMAGNETIC PROPERTIES

1. THEORETICAL TREATMENT

For phenomena involving matter and electromagnetic radiation, it is conventional to treat the interaction between the two parts as perturbation. In the usual "semiclassical treatment," the Maxwell equations are used for the radiation field, and the energy quantization, photon energy, of the field is invoked wherever necessary. Consider a crystalline solid with a perfect lattice and a uniform electron concentration; the fact that the electrons are discrete particles gives rise to Rayleigh–Tyndall scattering, but the effect is negligible for radiation having a wavelength much longer than the interelectron distance. Furthermore, confine considerations to phenomena that do not involve some radiation of modified frequency, resulting from the interaction of the prime radiation with matter as in various inelastic scatterings and photoluminescence. The matter is then characterized by two parameters, the dielectric constant $\tilde{\varepsilon}$ and the permeability $\tilde{\mu}$, which express the constitutive relations between the field vectors:

$$\mathbf{D}/\mathbf{E} = \tilde{\varepsilon} \qquad \text{and} \qquad \mathbf{B}/\mathbf{H} = \tilde{\mu}. \tag{6.62}$$

They are directly related to the electric and magnetic susceptibilities $\tilde{\chi}_e$ and $\tilde{\chi}_m$:

$$\tilde{\varepsilon} = 1 + 4\pi\tilde{\chi}_e \qquad \text{and} \qquad \tilde{\mu} = 1 + 4\pi\tilde{\chi}_m. \tag{6.63}$$

We shall consider materials for which only the dielectric constant is of significance. The parameter is a tensor quantity for crystalline solids. It reduces to a scalar quantity for crystals of cubic symmetry and for polycrystalline materials that may be considered isotropic. The dielectric constant, more precisely the dielectric function, pertaining to an electromagnetic field depends on the nature of the field. For a plane wave, it depends on the frequency ω, the wavevector $\boldsymbol{\kappa}$, and the polarization. For a specified type of polarization,

$$\tilde{\varepsilon}(\boldsymbol{\kappa},\omega) = \tilde{\varepsilon}_1(\boldsymbol{\kappa},\omega) + i\tilde{\varepsilon}_2(\boldsymbol{\kappa},\omega) = \tilde{\varepsilon}_1(\boldsymbol{\kappa},\omega) + i4\pi\tilde{\sigma}(\boldsymbol{\kappa},\omega)/\omega, \tag{6.64}$$

where $\tilde{\sigma}$ is the conductivity, a real quantity. The last term is taken with the plus sign under the assumption that the time dependence of the field is represented by $\exp(-i\omega t)$. The complex dielectric constant may be replaced by a complex conductivity $\tilde{\sigma}_c$:

$$\tilde{\varepsilon}(\boldsymbol{\kappa},\omega) = 1 + i4\pi\tilde{\sigma}_c(\boldsymbol{\kappa},\omega)/\omega; \tag{6.65}$$

$\tilde{\sigma}_c$ is defined by

$$\mathbf{J}(\boldsymbol{\kappa},\omega) = \tilde{\sigma}_c(\boldsymbol{\kappa},\omega)\mathbf{E}(\boldsymbol{\kappa},\omega),$$

where $\mathbf{J}$ is the electric current density.

For optical radiation and radiation of longer wavelength, the wavevector is negligibly small in the scale of the Brillouin zone of a crystal, and we may consider $\kappa = 0$ and sometimes omit $\boldsymbol{\kappa}$ in specifying $\tilde{\varepsilon}$.

A field vector of a plane wave with a wave-normal $\hat{\boldsymbol{\kappa}}$ is proportional to

$$\exp[i(\boldsymbol{\kappa}\cdot\mathbf{r} - \omega t)] \equiv \exp\left[i\omega\left(\frac{N}{c}\hat{\boldsymbol{\kappa}}\cdot\mathbf{r} - t\right)\right].$$

The complex refractive index

$$N = n + ik \tag{6.66}$$

consists of an ordinary refractive index n and an extinction coefficient k. N is related to $\tilde{\varepsilon}$ and $\tilde{\mu}$ of the medium:

$$N^2\mathbf{E} - N^2\hat{\kappa}(\hat{\kappa} \cdot \mathbf{E}) = \tilde{\mu}\tilde{\varepsilon}\mathbf{E}. \tag{6.67}$$

For a general crystal, wave analysis is rather involved. We shall confine our consideration to nonmagnetic, $\mu = 1$, crystals of sufficiently high symmetries, such that the principal axes of ϵ_1 and ϵ_2 coincide. Then

$$N_j = (\varepsilon_{1j} + \varepsilon_{2j})^{1/2} \tag{6.68}$$

for each of the principal axes, $j = x, y, z$. For a wave-normal $\hat{\kappa}$, N is determined by the Fresnel equation

$$\frac{\kappa_x^2}{N^{-2} - N_x^{-2}} + \frac{\kappa_y^2}{N^{-2} - N_y^{-2}} + \frac{\kappa_z^2}{N^{-2} - N_z^{-2}} = 0. \tag{6.69}$$

The equation gives two solutions of N for each wave-normal $\hat{\kappa}$. The two principal vibrations with electrical displacement vectors $\mathbf{D}'$ and $\mathbf{D}''$ are normal to each other, that is, $\mathbf{D}' \cdot \mathbf{D}'' = 0$. For each vibration, the ratios $D_x : D_y : D_z$ are in general complex; that is, the vibration is elliptical rather than linear. Furthermore, the electrical displacement vector may be not perpendicular to $\hat{\kappa}$. The two waves differ with respect to refraction, giving rise to the phenomenon of birefringence. Their differences in attenuation and in frequency dependence of attenuation are responsible for the phenomenon of pleochroism.

A useful parameter is the reflecting power R, which is the fraction of radiation reflected from a semiinfinite specimen of the material, for normal incidence of radiation in vacuo. Provided the bulk property of the material is maintained up to the surface, the reflecting power is given by

$$R = \frac{(n-1)^2 + k^2}{(n+1)^2 + k^2}. \tag{6.70}$$

The dissipation rate W of the field energy per unit volume is related to the conductivity $\tilde{\sigma}$:

$$W/2E_0^2 = \hat{E}_0 \cdot \tilde{\sigma} \cdot \hat{E}_0, \tag{6.71}$$

where $2\mathbf{E}_0$ is the amplitude of the electric field. The intensity of the wave attenuates exponentially, $\exp(-\alpha x)$, in its propagation along a principal axis, with an absorption coefficient

$$\alpha = 2k\omega/c = 4\pi\sigma/cn. \tag{6.72}$$

The two parts, $\tilde{\varepsilon}_1(\omega)$ and $\tilde{\varepsilon}_2(\omega)$, of the complex dielectric constant are related by the Kramers–Kronig relations:

$$\tilde{\varepsilon}_1(\omega) - 1 = \frac{2}{\pi} P \int_0^\infty \omega' \tilde{\varepsilon}_2(\omega')(\omega'^2 - \omega^2)^{-1} \, d\omega',$$

and

$$\tilde{\varepsilon}_2(\omega) - \frac{4\pi\sigma_0}{\omega} = \frac{2\omega}{\pi} P \int_0^\infty \tilde{\varepsilon}_1(\omega')(\omega^2 - \omega'^2)^{-1} \, d\omega'. \tag{6.73}$$

σ_0 is the dc conductivity, and P denotes the principal value. These relations are a direct consequence of causality. They are subject only to the conditions that the response of matter to field is linear and that the quantities on the left-hand side are bounded. Similarly, the two parts, $n(\omega)$ and $k(\omega)$, of the complex index of refraction are related:

$$n(\omega) - 1 = \frac{2}{\pi} P \int_0^\infty \omega' k(\omega')(\omega'^2 - \omega^2)^{-1} \, d\omega'. \tag{6.74}$$

a. Interaction of Matter with Radiation

For a system of radiation field and matter, the Hamiltonian of the entire system should be considered as a whole. In the semiclassical treatment, the Hamiltonian is taken to be

$$H = H_0 + H_\mathrm{I}, \tag{6.75}$$

where H_0 is the Hamiltonian of the matter in the absence of radiation field and H_I represents the interaction of the matter with a specified radiation. In treating H_I as a perturbation, the electromagnetic parameters that pertain to the equilibrium state corresponding to H_0 can be calculated.

Consider a group of charged particles of charge e and mass m. Except for a weak effect of an intrinsic magnetic moment that may be possessed by the particles, the interaction with radiation is given by

$$H_\mathrm{I} = \sum_i \left[\frac{e}{mc} \mathbf{p}_i \cdot \mathbf{A}(\mathbf{r}_i) + \frac{ie\hbar}{2mc} \nabla \cdot \mathbf{A} + \frac{e^2}{2mc^2} A^2 \right], \tag{6.76}$$

where i is the index of particles and $\mathbf{A}$ is the vector potential of radiation. The term in A^2 may be neglected as a small quantity of higher order. The matrix element of H_{I} connecting two eigenstates, $|S\rangle$ and $|S'\rangle$, of H_0 is given by

$$\langle S'|H_{\mathrm{I}}|S\rangle = \frac{e}{i\omega}\mathbf{E}\cdot(\mathbf{j}^{+}_{S'S}e^{-i\omega t} - \mathbf{j}^{-}_{S'S}e^{i\omega t}), \tag{6.77}$$

where $\mathbf{j}$ is the operator

$$\mathbf{j}^{\pm} = \sum_i \left(\frac{\mathbf{p}_i}{m} + \frac{\hbar\boldsymbol{\kappa}}{2m}\right)\exp(\pm i\boldsymbol{\kappa}\cdot\mathbf{r}_i). \tag{6.78}$$

According to first-order perturbation theory, the transition from $|S\rangle$ to $|S'\rangle$ occurs at a rate

$$T_{S'S} = \frac{2\pi}{\hbar}|\langle S'|H_{\mathrm{I}}|S\rangle|^2\delta(E_{S'} - E_S \mp \hbar\omega). \tag{6.79}$$

The upper signs apply to a transition with the absorption of a photon, and the lower signs apply to emission of a photon. The δ function expresses the energy conservation of the entire system. The transition increases the energy of the matter at a rate $\pm\hbar\omega T_{S'S}$.

For a system in the one-particle approximation, the rate of transitions involving state i as the initial state is given by

$$\begin{aligned} T_i &= T_i^{\mathrm{a}} + T_i^{\mathrm{e}} \\ &= \frac{2\pi}{\hbar}\sum_{i'} f_i f'_{i'}|\langle i'|H_{\mathrm{I}}|i\rangle|^2\delta(\epsilon_{i'} - \epsilon_i - \hbar\omega) \\ &\quad + \frac{2\pi}{\hbar}\sum_{i''} f_i f'_{i''}|\langle i''|H_{\mathrm{I}}|i\rangle|^2\delta(\epsilon_{i''} - \epsilon_i + \hbar\omega), \end{aligned} \tag{6.80}$$

where T_i^{a} stands for transitions with photon absorption and T_i^{e} stands for transitions with photon emission. f_i is the distribution function of state i.

$f' = 1$ for a system of bosons,

$f' = 1 - f$ for a system of fermions.

The statistical rate of energy absorption by the system is

$$W = \sum_i \hbar\omega(T_i^{\mathrm{a}} - T_i^{\mathrm{e}}), \tag{6.81}$$

which determines the real conductivity $\tilde{\sigma}$ according to (6.71). If the states i are Bloch states of electrons, the matrix elements of $\mathbf{p}$

and j have the following selection rules, known as the conditions of wavevector conservation:

$$\begin{aligned} \langle m'\mathbf{k}'|\mathbf{p}|m\mathbf{k}\rangle &= 0 \qquad \text{for} \quad \mathbf{k}' \neq \mathbf{k}, \\ \langle m'\mathbf{k}'|\mathbf{j}^{+}|m\mathbf{k}\rangle &= 0 \qquad \text{for} \quad \mathbf{k}' \neq \mathbf{k}+\boldsymbol{\kappa}, \\ \langle m'\mathbf{k}'|\mathbf{j}^{-}|m\mathbf{k}\rangle &= 0 \qquad \text{for} \quad \mathbf{k}' \neq \mathbf{k}-\boldsymbol{\kappa}, \end{aligned} \tag{6.82}$$

where m is the index of the energy band and $\mathbf{k}$ is the wavevector.

The complex conductivity $\tilde{\sigma}_c$ or the complex dielectric constant $\tilde{\varepsilon}$ represents the relationship between the radiation field and the induced electric current density. The expectation value of the induced current density is given by

$$\langle\psi|\mathbf{J}|\psi\rangle - \langle\psi_0|\mathbf{J}|\psi_0\rangle,$$

where ψ_0 is the unperturbed wavefunction of charged particles, ψ is the wavefunction perturbed by the radiation, and $\mathbf{J}$ is the operator

$$\mathbf{J} = \sum_{i=1}^{N} \frac{e}{m}\mathbf{p}_i$$

for the particles of concentration N. The statistical value of $\langle\psi_0|\mathbf{J}|\psi_0\rangle$ is zero. The induced current density, the statistical value of $\langle\psi|\mathbf{J}|\psi\rangle$, can be calculated by using ψ given by the first-order perturbation theory. The expression obtained for the dielectric function is

$$\begin{aligned} \tilde{\varepsilon}(\boldsymbol{\kappa},\omega) = &\left(1 - \frac{4\pi e^2 N}{m\omega^2}\right)\tilde{1} \\ &+ \frac{4\pi e^2}{\hbar\omega^2}\left\{\sum_{\nu}\left[\frac{\mathbf{j}_{0\nu}^{+}(\boldsymbol{\kappa})\mathbf{j}_{\nu 0}(\boldsymbol{\kappa})}{\omega-\omega_{\nu 0}+iS} - \frac{\mathbf{j}_{0\nu}(\boldsymbol{\kappa})\mathbf{j}_{\nu 0}^{+}(\boldsymbol{\kappa})}{\omega+\omega_{\nu 0}+iS}\right]\right\}_{S\to 0}. \end{aligned} \tag{6.83}$$

The subscript 0 denotes the unperturbed state, and ν is an index for the various other states of the particle system. The factor S is introduced by the addition of $\exp(St)$ to the vector potential $\mathbf{A}$. This procedure ensures that the field is switched on adiabatically so that the response of the system is causal. In view of the fact that

$$\lim_{S\to 0}\frac{1}{x+iS} = P\frac{1}{x} - i\pi\delta(x), \tag{6.84}$$

where P denotes the principal value, the introduction of S gives an imaginary part of the dielectric constant, in accordance with the Kramers–Kronig relations. The imaginary part represents the conductivity, which can also be obtained by calculating the energy absorption W from the field according to Eq. (6.71). Without the introduction of S, we would get an induced current completely out of phase with the electric field $\mathbf{E}$, giving an imaginary conductivity or a real dielectric function $\tilde{\varepsilon} = \tilde{\varepsilon}_1$.

If the dielectric constant is isotropic, that is, if $\tilde{\varepsilon}$ is a scalar, the radiation field in the absence of space charge is purely transverse with $\mathbf{E} \perp \boldsymbol{\kappa}$. Therefore, the dielectric constant involved in the electromagnetic properties is often referred to as the transverse dielectric constant. The term with $\nabla \cdot \mathbf{A}$ in the perturbation H_{I} vanishes, and we get

$$\varepsilon(\boldsymbol{\kappa},\omega) = 1 - \frac{4\pi e N}{\hbar\omega^2} + \frac{4\pi e^2}{\hbar\omega^2} \sum_{\nu}{}' |\hat{E}\cdot\mathbf{j}_{\nu 0}(\boldsymbol{\kappa})|^2 \times \left\{ P\frac{2\omega_{\nu 0}}{\omega_{\nu 0}^2 - \omega^2} - i\pi\delta(\omega_{\nu 0} - \omega)\right\}. \tag{6.85}$$

It is interesting to note that

$$\varepsilon(0,\omega) = \varepsilon_{\parallel}(0,\omega) \tag{6.86}$$

in isotropic matter. For such a substance, the electromagnetic properties can be related to phenomena involving $\varepsilon_{\parallel}$, such as plasma oscillations and excitations produced by the passage of fast charged particles.

In the one-electron approximation, the contribution of electrons of a crystalline solid to the dielectric function has the expression

$$\tilde{\varepsilon}(0,\omega) = \left(1 - \frac{4\pi e^2 N}{m\omega^2}\right)\tilde{1} + \left\{ \frac{4\pi e^2}{m^2\omega^2\hbar} \sum_{mm'} \frac{(m\mathbf{k}|\mathbf{p}|m'\mathbf{k})(m'\mathbf{k}|\mathbf{p}|m\mathbf{k})}{\omega - \omega_{m'\mathbf{k},m\mathbf{k}} + iS} \times [f(\epsilon_{m'\mathbf{k}}) - f(\epsilon_{m\mathbf{k}})] \right\}_{S\to 0}, \tag{6.87}$$

where f is the electron distribution function in the unperturbed system. The tensor of oscillator strength introduced by (3.39) and (3.40) may be written as

$$f_{\mathbf{k},m'm} = \frac{2}{m}\frac{\mathbf{p}_{\mathbf{k},m'm}\mathbf{p}_{\mathbf{k},m'm}}{\hbar\omega_{\mathbf{k},m'm}}, \tag{6.88}$$

in terms of which the reciprocal-effective-mass tensor has the form

$$\left(\widetilde{\frac{1}{m^*}}\right)_{m,\mathbf{k}} = \frac{1}{m}\left[\tilde{1} - \frac{1}{2}\sum_{m'\neq m}(\tilde{f}_{\mathbf{k},m'm} - \tilde{f}_{\mathbf{k},mm'})\right]. \tag{6.89}$$

Introducing the oscillator strengths and the reciprocal effective mass into the expression of $\epsilon(0,\omega)$, we get

$$\begin{aligned}\tilde{\varepsilon}_1(0,\omega) = 1 &- \frac{4\pi e^2}{\omega^2}\sum_{\mathbf{k}m} f(\epsilon_{m\mathbf{k}})\left(\frac{1}{m^*}\right)_{m\mathbf{k}} \\ &+ \frac{4\pi e^2}{2m\omega}\sum_{\mathbf{k}mm'}{}' f(\epsilon_{m\mathbf{k}})\left(\frac{\tilde{f}_{\mathbf{k},m'm}}{\omega_{\mathbf{k},m'm} - \omega} - \frac{\tilde{f}_{\mathbf{k},mm'}}{\omega_{\mathbf{k},mm'} - \omega}\right)\end{aligned} \tag{6.90}$$

and

$$\tilde{\varepsilon}_2(0,\omega) = 4\pi\frac{\pi e^2}{2m\omega}\sum_{\mathbf{k}mm'}{}'[f(\epsilon_{m\mathbf{k}}) - f(\epsilon_{m'\mathbf{k}})]\tilde{f}_{\mathbf{k},m'm}\delta(\omega_{\mathbf{k},m'm} - \omega). \tag{6.91}$$

The expression of $\tilde{\varepsilon}_2$ and the last term of $\tilde{\varepsilon}_1$ represent the interband effect that involves oscillator strengths connecting various energy bands. Two energy bands m and m' each fully occupied by electrons do not contribute to $\tilde{\varepsilon}_1$ and $\tilde{\varepsilon}_2$ by interband transitions. In order for a pair of energy bands to give an interband effect, it is necessary for some states to be occupied in one band while states of the same **k**'s are empty in the other band. Furthermore, the existence of the interband effect in $\tilde{\varepsilon}_2$ requires that the difference $\hbar\omega_{\mathbf{k},mm'}$ in electron energy be equal to the photon energy $\hbar\omega$ of the radiation. At $\omega = \omega_{\mathbf{k},mm'}$, the absorption of radiation increases sharply, being referred to as an absorption edge.

The second term of $\tilde{\varepsilon}_1$ gives contributions of individual energy bands, and it may be called the intraband effect. It is given by free carriers, since the summation of $1/m^*$ over an energy band fully occupied by electrons, $f(\epsilon_{\mathbf{k}}) = 1$ for all **k**, gives zero. As to $\tilde{\varepsilon}_2$, the existence of free carriers can influence the interband effect only through the distribution function $f(\epsilon_{\mathbf{k}})$ of the particular band. There is no intraband term at all. As in the case of dc conduction, intraband processes of free carriers can produce energy absorption from the field only if some mechanism of carrier scattering is involved.

For an isotropic substance, we have

$$f_{\mathbf{k},m'm} = -f_{\mathbf{k},mm'} = \frac{2}{m}\frac{|(\hat{E}\cdot\mathbf{p})_{\mathbf{k},m'm}|^2}{\hbar\omega_{\mathbf{k},m'm}} = \frac{2}{3m}\frac{|\mathbf{p}_{\mathbf{k},m'm}|^2}{\hbar\omega_{\mathbf{k},m'm}} \tag{6.92}$$

and

$$\varepsilon_1 = 1 - \frac{4\pi e^2}{\omega^2} \sum_{m\mathbf{k}} f(\epsilon_{m\mathbf{k}}) \left(\frac{1}{m^*}\right)_{m\mathbf{k}} + \frac{4\pi e^2}{m} \sum_{\mathbf{k}mm'}{}' f(\epsilon_{m\mathbf{k}}) \frac{f_{\mathbf{k},m'm}}{\omega^2_{\mathbf{k},m'm} - \omega^2}, \tag{6.93}$$

$$\varepsilon_2 = \frac{4\pi}{\omega} \frac{e^2}{2m} \sum_{\mathbf{k}mm'}{}' [f(\epsilon_{m\mathbf{k}}) - f(\epsilon_{m'\mathbf{k}})] f_{\mathbf{k},m'm} \delta(\omega_{\mathbf{k},m'm} - \omega). \tag{6.94}$$

Lattice Imperfections. Imperfections in a crystal that are sufficiently separated from each other introduce localized electronic states and local modes of lattice vibration that are characteristic of the type of imperfection. The effect on the electromagnetic properties of the crystal can be obtained by summing the results calculated for the individual imperfection centers.

b. Inherent Interactions in the Crystal

So far, we have considered the crystal in the one-electron approximation with a stationary lattice. In this approximation, the perturbation H_{I} of radiation produces transitions between two states of the crystal, $|0\rangle$ to $|F\rangle$, at a rate given by (6.79). Various interactions H_0', such as electron–electron interaction and electron interaction with lattice vibration, have been neglected. Consider now that H_0' together with an H_{I} may in the second-order perturbation produce a $|0\rangle \to |F\rangle$ transition via each of two intermediate states $|I\rangle$ and $|I'\rangle$. The rate of transition is given by

$$P_{f0} = \frac{2\pi}{\hbar} \left| \frac{\langle F|H_0'|I\rangle\langle I|H_{\mathrm{I}}|0\rangle}{\mathcal{E}_0 - \mathcal{E}_i} + \frac{\langle F|H_{\mathrm{I}}|I'\rangle\langle I'|H_0'|0\rangle}{\mathcal{E}_0 - \mathcal{E}_{i'}} \right|^2 \delta(\mathcal{E}_f - \mathcal{E}_0), \tag{6.95}$$

where $\mathcal{E}$ is the energy of the system, crystal and radiation, and the subscripts 0, i, i', and f indicate the states of the system. Calculations may have to be carried to higher orders of perturbation if some H_0' has a large effect. In principle, it would be desirable to first take into account all H_0''s for the crystal before treating the perturbation H_{I} of radiation.

Transitions involving some H_0' are particularly important when $\langle F|H_{\mathrm{I}}|0\rangle = 0$. Sometimes the effect of H_0' is treated approximately as a smearing of the states $|F\rangle$ and $|0\rangle$. Such an approach may be justifiable if $\langle F|H_{\mathrm{I}}|0\rangle \neq 0$ and the smearing is small compared with the energy difference $|E_{\mathrm{F}} - E_0|$ of the crystal.

Inelastic Scattering. The real part ε_1 of the dielectric function determines the refraction or elastic scattering of the radiation. The inelastic scattering involves a change in the frequency of the radiation. The treatment of elastic and inelastic scattering presented in Section II.5 is limited to the consideration of a lattice. Radiation scattering involves the interaction $H_{\rm I}$ of the matter with the primary radiation and the interaction $H_{\rm I'}$ of the matter with the scattered radiation. The frequencies of the two radiations are the same, $\omega' = \omega$, in elastic scattering, whereas they are different, $\omega' \neq \omega$, in inelastic scattering. Elastic scattering is produced by second-order transitions, the transition matrix of which has the form

$$\langle F|H_{\rm I'}|I\rangle\langle I|H_{\rm I}|0\rangle = \langle 0|H_{\rm I'}|I\rangle \quad \text{or} \quad \langle 0|H_{\rm I}|I\rangle\langle I|H_{\rm I'}|0\rangle.$$

The matter is in the same state before and after the transition, that is, $|F\rangle = |0\rangle$. The virtual transition due to $H_{\rm I}$ involves absorption of a photon $\hbar\omega$ of the primary radiation, and that due to $H_{\rm I'}$ involves the emission of a photon $\hbar\omega' = \hbar\omega$ of the scattered radiation.

The inelastic scattering can be calculated in third-order perturbation theory, including an inherent interaction of the matter. The transition matrix has the form

$$\langle F|H_{\rm I'}|I_2\rangle\langle I_2|H_0'|I_1\rangle\langle I_1|H_{\rm I}|0\rangle, \tag{6.96}$$

with the option of interchanging $H_{\rm I}$, $H_{\rm I'}$, and H_0'. Such a transition changes the state of the matter, with a change ΔE in energy. It follows from energy conservation of the system that ω' differs from ω, according to

$$\hbar\omega' - \hbar\omega = -\Delta E. \tag{6.97}$$

The difference $\omega' - \omega$ is referred to as the Stokes shift or the anti-Stokes shift, depending on whether it is negative or positive. If ΔE is quantized, the intensity of the scattered radiation as a function of ω' shows a number of characteristic peaks.

2. LOW-FREQUENCY PHENOMENA

Strong absorption due to interband transitions of electrons begins at some frequency. Phenomena occurring at lower frequencies may be produced by (1) intraband transitions of free carriers, (2) electronic transitions involving localized states of imperfection, or (3) the effects of lattice vibration.

a. Intraband Effect of Free Carriers

It has been pointed out in connection with (6.91) that the intraband effect of free carriers shows up in ε_1 but not in ε_2. ε_2 represents the absorption of radiation energy by matter. The matrix element $(m\mathbf{k}|\mathbf{p}|m'\mathbf{k})$ of perturbation H_{I} requires wavevector conservation, Eq. (6.82). For absorption to occur, the condition $\omega_{\mathbf{k},m'm}=\omega$ of energy conservation is also required. The impossibility of satisfying both requirements simultaneously eliminates the intraband effect of free carriers from ε_2. This obstacle may not be present if interactions H_0' in the crystal are considered together with H_{I} in a second-order perturbation discussion in Section VI.C.1.a; energy conservation is not required for virtual transitions produced by H_{I}. The Drude–Kronig theory considers electrons of mass m^* moving under an applied ac field, with a scattering frequency $1/\tau$ representing the effect of H_0'. The expression obtained for the complex conductivity is

$$\sigma_{\text{c}}^{\text{f}}=\frac{Ne^2}{m^*}\frac{1}{1-i\omega\tau}, \tag{6.98}$$

where N is the concentration of charge carriers. The expression gives

$$\varepsilon_1^{\text{f}}=-4\pi\frac{Ne^2\tau^2}{m^*}\frac{1}{1+(\omega\tau)^2}$$

and

$$\sigma^{\text{f}}=\frac{Ne^2\tau}{m^*}\frac{1}{1+(\omega\tau)^2}.$$

In the absence of scattering, $1/\tau=0$, we get $\sigma^{\text{f}}=0$ and an ε_1^{f} equal to the second term of Eq. (6.90), the expression for free carriers in one energy band.

The use of a constant parameter $1/\tau$ to represent the effect of scattering is a rough approximation. The energy of a carrier is raised considerably in the absorption of a photon, and the increase depends on the frequency ω of radiation. The fact that electron scattering is energy-dependent affects the frequency dependence of absorption. When the modification of carrier energy by $\hbar\omega$ is significant, the proper calculation by second-order perturbation should be made. The Drude–Kronig theory gives

$$\sigma(\omega)=\frac{Ne^2}{m^*\tau}\omega^{-2} \tag{6.99}$$

for $\omega\tau > 1$. Different results are given by the more appropriate calculation that depend on the dominant mechanism of scattering. For example, consider the simple case of carriers having one scalar effective mass m^* and classical distribution. The scattering by acoustical modes of lattice vibration, Section V.1.a, leads to

$$\sigma(\omega) = N\frac{e^2(2m^*)^{1/2}}{3\pi\hbar^2}\frac{C^2kT}{dc_l^2}(\hbar\omega)^{-3/2}f(x), \tag{6.100}$$

where d is the density of the crystal, c_l is the sound velocity, C is the deformation potential,

$$f(x) = (2/\pi)^{1/2}x^{1/2}e^x(1-e^{-2x})K_2(x),$$
$$x = \hbar\omega/2kT,$$

and $K_2(x)$ is the modified Bessel function.

On the other hand, the scattering by centers of charge Ze, Section V.2, gives

$$\sigma(\omega) = NN_i\frac{8}{3}\left(\frac{Ze^2}{\varepsilon}\right)^2\frac{e^2\hbar^2}{m^*}(2\pi m^*kT)^{-1/2}(\hbar\omega)^{-3}(1-e^{-2x})e^xK_0(x), \tag{6.101}$$

where N_i is the concentration of scattering centers and $K_0(x)$ is the modified Bessel function of order zero.

The contribution $\varepsilon_1^{\mathrm{f}}$ of free carriers is important for the reflecting power R of the crystal. The effect of $\varepsilon_1^{\mathrm{f}}$ can easily be deduced by neglecting electron scattering. At low frequencies, the negative $\varepsilon_1^{\mathrm{f}}$ is large, making ε_1 negative, $n = 0$, and $R = 1$ of total reflection. As $\varepsilon_1^{\mathrm{f}}$ decreases in magnitude with increasing frequency, ε_1 becomes positive and n rises from zero, at some frequency ω_{p}. The change of n makes R decrease from unity at ω_{p}. At a frequency $\omega_m > \omega_{\mathrm{p}}$, $n = 1$ is reached, and we get $R = 0$. A further increase in frequency makes n exceed unity and R rise from zero. The variation in R due to the effect of free carriers is referred to as the plasma effect. The above discussion neglects the effect of the extinction coefficient k given by the absorption. Actually, the shape of the $R(\omega)$ spectra is affected by the absorption due to carrier scattering. In some semiconductors that show a weak scattering, a minimum of $R \sim 0$ has actually been observed.

b. Magneto-optics, Low Magnetic Field

For a crystal in equilibrium under an applied magnetic field, the response of free carriers to radiation shows the effect of the magnetic field even when the field is too weak to make the quantization of electronic energy levels significant. An applied field alters the symmetry of the crystal and thereby affects the symmetry nature of the properties. Consider, for example, a cubic crystal, for which the tensor of complex dielectric constant or the tensor of complex conductivity can be represented by a scalar. Let us use $\tilde{\sigma}$ for the intraband effect of free carriers and $\tilde{\varepsilon}$ to cover the other processes. With a magnetic field along one of the cubic axes, say $\hat{z}$, $\tilde{\varepsilon}$ and $\tilde{\sigma}$ each no longer has only diagonal elements that are equal to each other. Each of the tensor quantities then has five nonzero components: (xx), (yy), (zz), (xy), and $(yx) = -(xy)$.

According to (6.46), the components of dc conductivity are given by

$$\begin{aligned} \sigma_{zz} &= \frac{ne^2\tau}{m^*} = \frac{ne^2}{m^*\gamma} = \sigma_0, \\ \sigma_{xx} &= \sigma_{yy} = \sigma_0 \frac{1}{1+(\omega_c\tau)^2} = \frac{ne^2}{m^*}\frac{\gamma}{\gamma^2+\omega_c^2}, \end{aligned} \tag{6.102}$$

and

$$\sigma_{xy} = -\sigma_{yx} = \frac{ne^2}{m^*}\frac{\omega_c}{\gamma^2+\omega_c^2}$$

for free carriers of effective mass m^* and a constant relaxation time τ or scattering frequency γ, under an applied magnetic field $B \parallel \hat{z}$. In a radiation field varying as $\exp(-i\omega t)$, $\partial f/\partial t = \Delta f \cdot \exp(-i\omega t)$ in the Boltzmann equation, instead of $\partial f/\partial t = 0$. Hence γ is replaced by $\gamma - i\omega$. Consequently, we get

$$\begin{aligned} \sigma_{zz} &= \frac{ne^2}{m^*}\frac{1}{\gamma - i\omega}, \qquad \sigma_{xx} = \sigma_{yy} = \frac{ne^2}{m^*}\frac{\gamma - i\omega}{(\gamma - i\omega)^2+\omega_c^2}, \\ \sigma_{xy} &= -\sigma_{yx} = \frac{ne^2}{m^*}\frac{\omega_c}{(\gamma - i\omega)^2+\omega_c^2}. \end{aligned} \tag{6.103}$$

***Wave Propagation along* B, K‖B‖$\hat{z}$.** According to (6.67), $E_z = 0$, that is, $\mathbf{E} \perp \mathbf{B}$. With $\mathbf{E} = \mathbf{i}E_x + \mathbf{j}E_y$, we get

$$N^2 E_x = (\varepsilon_{xx} + i4\pi\sigma_{xx}/\omega)E_x + (\varepsilon_{xy} + i4\pi\sigma_{xy}/\omega)E_y$$

and

$$N^2 E_y = (\varepsilon_{xx} + i4\pi\sigma_{xx}/\omega)E_y - (\varepsilon_{xy} + i4\pi\sigma_{xy}/\omega)E_x. \tag{6.104}$$

These equations show that

$$E_x^2 = -E_y^2 \qquad \text{or} \qquad E_y = \pm iE_x.$$

In other words, the wave is circularly polarized. The sign, + or −, in the relation of E_y to E_x indicates the direction of rotation of **E**. The solutions of N for the two waves are

$$N_{\pm}^2 = (\varepsilon_{xx} + i4\pi\sigma_{xx}/\omega) \pm i(\varepsilon_{xy} + i4\pi\sigma_{xy}). \tag{6.105}$$

The superposition of two circularly polarized waves, + and −, gives a direction of **E** for each plane $\mathbf{r}\cdot\boldsymbol{\kappa}$. In the case of $N_+ = N_-$, the direction of **E** does not depend on $\mathbf{r}\cdot\boldsymbol{\kappa}$ and the wave is plane polarized. With $N_+ \neq N_-$, the direction of **E** rotates by a solid angle of

$$\theta = \frac{\omega}{2c}(N_+ - N_-) \tag{6.106}$$

as $\mathbf{r}\cdot\boldsymbol{\kappa}$ varies by unity. The so-called Faraday rotation has been investigated for the study of free carriers.

Under a high magnetic field such that

$$\omega_c = eB/m^*c \gg \omega \qquad \text{and} \qquad \omega_c \gg \gamma,$$

we get

$$\sigma_{xy} \sim \frac{ne^2}{m^*\omega_c} \gg \frac{ne^2}{m^*\omega_c}\frac{\gamma + i\omega}{\omega_c} \sim \sigma_{xx}. \tag{6.107}$$

If the concentration n of free carriers is sufficiently large, we also have

$$\frac{4\pi}{\omega}\sigma_{xy} \gg \varepsilon_{xx}, \varepsilon_{xy}.$$

Under these conditions,

$$N_{\pm}^2 \sim \pm\frac{4\pi}{\omega}\frac{ne^2}{m^*\omega_c}. \tag{6.108}$$

We see that only the circularly polarized wave with the + sign can propagate in the crystal. Such waves, so-called helicon waves, have been observed in some metals, with wavelengths ∼ 1 cm and $\omega \sim$ 10 Hz. The phase velocity is reduced to ∼ 10 cm/s corresponding to N of the order of 10^9.

***Wave Propagation Perpendicular to* B, $\boldsymbol{\kappa} \parallel \hat{y} \perp \mathbf{B} \parallel \hat{z}$.** According to (6.67),

$$\begin{aligned} N^2E_x &= (\varepsilon_{xx} + i4\pi\sigma_{xx}/\omega)E_x + (\varepsilon_{xy} + i4\pi\sigma_{xy}/\omega)E_y, \\ N^2E_y - N^2E_y = 0 &= (\varepsilon_{xx} + i4\pi\sigma_{xx}/\omega)E_y - (\varepsilon_{xy} + i4\pi\sigma_{xy}/\omega)E_x, \end{aligned} \tag{6.109}$$

and

$$N^2 E_z = (\varepsilon_{zz} + i4\pi\sigma_{zz}/\omega)E_z.$$

It follows from the second equation that E_x and E_y are either both equal to zero or both different from zero. We may have the following two kinds of waves:

1. $\mathbf{E} \perp \mathbf{B}$; $E_z = 0$, $E_x \neq 0$, $E_y \neq 0$.

$$N_\perp^2 = \left[\left(\varepsilon_{xx} + \frac{i4\pi\sigma_{xx}}{\omega}\right) + \left(\varepsilon_{xy} + \frac{i4\pi\sigma_{xy}}{\omega}\right)\right] \times \left[\left(\varepsilon_{xx} + \frac{i4\pi\sigma_{xx}}{\omega}\right)\left(\varepsilon_{xy} + \frac{i4\pi\sigma_{xy}}{\omega}\right)\right]^{-1}. \quad (6.110)$$

The direction of **E** rotates in the xy plane containing the wave-normal $\boldsymbol{\kappa}$ as the wave propagates.

2. $\mathbf{E} \parallel \mathbf{B}$; $E_x = E_y = 0$. The wave is transverse and linearly polarized.

$$N_\parallel^2 = \varepsilon_{zz} + i4\pi\sigma_{zz}/\omega. \quad (6.111)$$

The distinction between $N_\perp$ and $N_\parallel$ is known as the Voight effect. Along with the Faraday effect, the Voight effect has been used as a means to study the free carriers.

c. Cyclotron Resonance Absorption

For crystals under a sufficiently high applied magnetic field, consideration should be based on Landau levels discussed in Section IV.A.3. Electron transitions between subbands can occur by interaction with a radiation field. Consider the more general case treated in Section IV.A.3: The effective mass is a tensor and the magnetic field **B** may not be parallel to an axis of the tensor. From Eqs. (4.32) and (6.76), we see that the interaction H_I of an electron with radiation is given by

$$H_\mathrm{I} = \frac{ie\hbar}{cm}\left[A_x\frac{\partial}{\partial x} + A_y\left(\frac{\partial}{\partial y} - \frac{ieB}{c\hbar\mu_c}x - \frac{ieB}{c\hbar M}y\right) + A_z\frac{\partial}{\partial z}\right] \quad (6.112)$$

in the "reduced coordinate system" (x, y, z). A_x, A_y, A_z are the components of vector potential of the radiation. μ_c and M are defined in (4.33). Expression (6.112) omits

$$\frac{ie\hbar}{2mc}\left(\frac{1}{\mu_1}\frac{\partial A_1}{\partial x_1} + \frac{1}{\mu_2}\frac{\partial A_2}{\partial x_2} + \frac{1}{\mu_3}\frac{\partial A_3}{\partial x_3}\right),$$

which can be shown to have a negligible effect.

A state of an electron is characterized by n, k_y, k_z, where n is the quantum number of the subband. Note the following matrix elements between states ν and ν':

$$\left(\frac{\partial}{\partial y} - \frac{ieB}{c\hbar\mu_c}x - \frac{ieB}{c\hbar M}y\right)_{\nu'\nu} = i\left|\frac{\partial}{\partial x}\right|_{\nu'\nu} - i\left(\frac{eB}{2c\hbar\mu_c}\right)^{1/2}\delta_{k_y'k_y}\delta_{k_z'k_z} \times \begin{cases}(n+1)^{1/2}, & \text{for } n' = n+1 \\ n^{1/2}, & \text{for } n' = n-1\end{cases} \tag{6.113}$$

and

$$\left(\frac{\partial}{\partial z}\right)_{\nu'\nu} = ik_z\delta_{k_y'k_y}\delta_{k_z'k_z}\delta_{n'n};$$

the wavevector $\boldsymbol{\kappa}$ of radiation is neglected, since it is small in the scale of $\mathbf{k}$. Consider polarized radiation propagating perpendicular to the applied magnetic field. Either (a) the wave is transverse with $\mathbf{A} \parallel \mathbf{B}$ or (b) the wave has $\mathbf{A} \perp \mathbf{B}$ and may not be purely transverse. The matrix elements for electron transition are given by

$$|(H_{\mathrm{I}})_{\nu'\nu}|^2 = \left(\frac{e\hbar}{mc}\right)^2\left[(A_x^2 + A_y^2)\left|\left(\frac{\partial}{\partial x}\right)_{\nu'\nu}\right|^2 + A_z^2\left|\left(\frac{\partial}{\partial z}\right)_{\nu'\nu}\right|^2\right] \quad \text{for case (a)}$$

and

$$|(H_{\mathrm{I}})_{\nu'\nu}|^2 = \left(\frac{e\hbar}{mc}\right)^2(A_x^2 + A_y^2)\left|\left(\frac{\partial}{\partial x}\right)_{\nu'\nu}\right|^2 \quad \text{for case (b).} \tag{6.114}$$

In view of (6.113), intersubband transitions can occur only with $n' - n = \pm 1$, which, since $k_z' = k_z$, corresponds to a change of

$$\Delta\epsilon = \epsilon_{n'} - \epsilon_n = \pm\hbar\frac{|eB|}{cm\mu_c} \tag{6.115}$$

in the electron energy. The condition of energy conservation requires $\Delta\epsilon$ to be the negative of the change in radiation energy, which is $-\hbar\omega$ for absorption and $+\hbar\omega$ for emission of a photon. Hence, electron transitions from n to $n+1$, $\Delta\epsilon > 0$, are associated with photon absorption, and transitions from n to $n-1$, $\Delta\epsilon < 0$, are associated with photon emission. In either case, we have

$$\omega = \frac{|e|}{cm\mu_c} B. \tag{6.116}$$

Photon absorption is usually the major phenomenon. With one of the two quantities, ω and B, kept constant, absorption occurs at a particular value of the other. The effect is cyclotron resonance absorption. It gives information on the effective mass of carriers.

Phonon–Cyclotron Resonance. Electron–phonon interaction H_0' leads to second-order intersubband transitions of electrons. Such a transition involves absorption or emission of a phonon in addition to absorption or emission of a photon. For the absorption of a photon, the rate of such a transition is

$$P \propto \left| \frac{\langle n'k_z', N_\mathbf{q} \pm 1 | H_0' | n''k_z, N_\mathbf{q}\rangle \langle n''k_z, N_\mathbf{q} | H_\mathrm{I} | nk_z, N_\mathbf{q}\rangle}{(\epsilon_{n''k_z'} - \epsilon_{nk_z}) - \hbar\omega} + \frac{\langle n'k_z', N_\mathbf{q} \pm 1 | H_\mathrm{I} | n'''k_z', N_\mathbf{q} \pm 1\rangle \langle n'''k_z', N_\mathbf{q} \pm 1 | H_0' | nk_z, N_\mathbf{q}\rangle}{\pm\hbar\omega_\mathbf{q} - \hbar\omega} \right|^2. \tag{6.117}$$

In the transition, an electron goes from state $|nk_z\rangle$ to state $|n'k_z'\rangle$, and a phonon of wavevector $\mathbf{q}$ is emitted (+) or absorbed (−). For the product of matrix elements of H_I and H_0' to be nonzero, the wavevectors have to satisfy the condition

$$k_z \mp q_z = 0 \qquad \text{and} \qquad k_x \mp q_x = k_x'.$$

The condition of energy conservation required for the transition is

$$\epsilon_{n'k_z'} - \epsilon_{nk_z} \pm \hbar\omega_\mathbf{q} = \hbar\omega,$$

where $\hbar\omega_\mathbf{q}$ is the energy of the participating phonon. Using (4.33), we get

$$\hbar\omega - \hbar\omega_\mathbf{q} = (n' - n)\frac{\hbar|eB|}{cm\mu_c} - \frac{\hbar^2}{2m}(q_z^2 - 2k_z q_z) \tag{6.118}$$

for transitions involving the emission of a phonon. Consider polar optical phonons, which are of special interest. Since the density of states in a subband increases toward a singularity at $k_z = 0$, as shown by (4.34), the transitions can be expected to peak around $q_z \sim 0 \sim k_z$. Optical phonons have finite energies that do not vary significantly with $\mathbf{q}$ in the range of small $\mathbf{q}$, and polar modes can interact strongly with electrons as shown in Section V.1.b. Absorption peaks occurring at

$$\hbar\omega - \hbar\omega_0 \simeq (n' - n)\frac{\hbar|eB|}{cm\mu_c} \tag{6.119}$$

are due to phonon–cyclotron resonance absorption. Because of the phonon involvement, n' is not limited to $n+1$. For a fixed ω or B, a series of resonances occur as the other one is varied. The phenomenon provides a means of investigating the relation among subbands, not limited to two neighboring bands. It also gives information on the optical phonon with regard to its energy $\hbar\omega_0$ and its interaction with the carriers.

d. Effect of Localized States

Localized states introduced by an imperfection center create the possibility of electron transition between two localized states or between a localized state and the states of energy bands. Some general remarks can be made concerning "shallow levels." For simplicity, consider the case where the nearest energy band is nondegenerate at the edge. From Section IV.A.2, we have the matrix element for electron transition involving a localized state $|i\rangle$:

$$\int \psi_j^* \mathbf{p} \psi_i \, d\mathbf{r} = \sum_{\mathbf{k},\mathbf{k}'} A_{\mathbf{k}'j}^* A_{\mathbf{k}i} \int \varphi_{\mathbf{k}'}^* \mathbf{p} \varphi_{\mathbf{k}} \, d\mathbf{r} = \sum_{\mathbf{k}} A_{\mathbf{k}j}^* A_{\mathbf{k}i} \frac{m}{\hbar} \nabla_{\mathbf{k}} \epsilon,$$

where $|j\rangle$ is either a state of the energy band or another shallow localized state of the imperfection center. The component along an axis α of the effective mass tensor is

$$\int \psi_j^* p_\alpha \psi_i \, d\mathbf{r} = \frac{m}{m_\alpha} \sum_{\mathbf{k}} A_{\mathbf{k}j}^* A_{\mathbf{k}i} \hbar k = \frac{m}{m_\alpha} \int F_j^*(\mathbf{r}) p_\alpha F_i(\mathbf{r}) \, d\mathbf{r}. \tag{6.120}$$

We see that the optical transitions of electrons are determined by the matrix elements of the envelope function $F(\mathbf{r})$.

Effect of Electron–Lattice Interaction. We have here the complication that the spatial geometry of the imperfection center and the surrounding ions may depend on electron occupation of the localized states. The complication applies also to the lattice vibrations of significance.

Let $\langle a\alpha|H_{\mathrm{I}}|b\beta\rangle$ denote the matrix element for a transition induced by H_{I}; a or b refers to the one-electron state involved in the transition, and α of β refers to the lattice. Consider the case where radiation perturbs the crystal only through its interaction with electrons. Then

$$\langle a\alpha|H_{\mathrm{I}}|b\beta\rangle = \int d\mathbf{R}\,\chi^*_{a\alpha}\chi_{b\beta}\int d\mathbf{r}\,\varphi^*_a(\mathbf{r},\mathbf{R})H_{\mathrm{I}}\varphi_b(\mathbf{r},\mathbf{R})$$
$$= (H_{\mathrm{I}})_{ab}\int d\mathbf{R}\,\chi^*_{a\alpha}\chi_{b\beta}. \tag{6.121}$$

Since the electronic function $\varphi(\mathbf{r},\mathbf{R})$ of the adiabatic approximation contains the coordinates $\mathbf{R}$ of ions as parameters, the quantity $(H_{\mathrm{I}})_{ab}$ obviously depends on $\chi_{a\alpha}$ and $\chi_{b\beta}$. It is customary to make the approximation of taking it to be independent of α and β.

In Eq. (2.5) for the lattice function $\chi(\mathbf{R})$, the contribution ε of an electron in the pertinent state is separated out from $E_{\mathbf{R}}$; $\varepsilon_a(\mathbf{R})$ for $\chi_{a\alpha}$ and $\varepsilon_b(\mathbf{R})$ for $\chi_{b\beta}$. Let $\mathbf{R}_0$ denote the equilibrium positions of the ions if $E_{\mathbf{R}}$ is replaced by $E_{\mathbf{R}} - \varepsilon$. ε can be considered a sum of two components:

$$\varepsilon = \varepsilon(\mathbf{R}_0) + \Delta\varepsilon(\mathbf{R} - \mathbf{R}_0). \tag{6.122}$$

Use the linear approximation, taking $\Delta\varepsilon$ to be proportional to $\mathbf{R} - \mathbf{R}_0$. Then $\Delta\varepsilon$ can be expressed in the following way:

$$\Delta\varepsilon = N^{-1/2}\sum_{j=1}^{N} A_j q_j,$$

where q_j stands for the normal coordinates of vibration mode j and A_j is the coupling coefficient of the electron and q_j. Consequently,

$$\hbar\omega_{ba} \equiv \varepsilon_b(\mathbf{R}_b) - \varepsilon_a(\mathbf{R}_a)$$
$$= \varepsilon_b(\mathbf{R}_0) - \varepsilon_a(\mathbf{R}_0) - \frac{1}{2N}\sum_{j=1}^{N}\left[\left(\frac{A_j^b}{\omega_j^b}\right)^2 - \left(\frac{A_j^a}{\omega_j^a}\right)^2\right]. \tag{6.123}$$

The last term represents the effect of the shift of equilibrium coordinates from $\mathbf{R}_0$ to $\mathbf{R}_a$ and $\mathbf{R}_b$, respectively, for the two electronic states a and b. The energy difference between the two states $|b\beta\rangle$ and $|a\alpha\rangle$ of the crystal is

$$E_{b\beta} - E_{a\alpha} = \hbar\omega_{ba} + \sum_{j=1}^{N}\left[(n_j^b + \tfrac{1}{2})\hbar\omega_j^b - (n_j^a + \tfrac{1}{2})\hbar\omega_j^a\right], \tag{6.124}$$

in the usual approximation. n_j and ω_j are the quantum number and frequency, respectively, of vibration mode j.

The transition probability from states containing φ_a to states with φ_b is determined by

$$I_{ba}(\hbar\omega) = |(H_{\mathrm{I}})_{ba}|^2 \mathrm{Av}_\alpha \sum_\beta \int d\mathbf{R} |\chi^*_{b\beta}\chi_{a\alpha}|^2 \delta(E_{b\beta} - E_{a\alpha} \mp \hbar\omega); \tag{6.125}$$

the + and − signs apply for the emission and absorption, respectively, of a photon. The expression takes into account final states for all the various β's of the lattice for individual α; the symbol Av_α stands for averaging statistically over the occupied initial vibrational states. The spectral distribution of transitions $a \to b$ is given by the frequency dependence of

$$G_{ba}(\hbar\omega) = \frac{I_{ba}(\hbar\omega)}{|(H_{\mathrm{I}})_{ba}|^2}. \tag{6.126}$$

The spectrum may be characterized by its various moments.

$$\langle \hbar\omega \rangle = \int (\hbar\omega) G(\hbar\omega)\, d(\hbar\omega) \tag{6.127}$$

gives the center of gravity of the spectrum, and

$$\langle (\hbar\omega - \langle \hbar\omega \rangle)^2 \rangle = \int (\hbar\omega - \langle \hbar\omega \rangle)^2 G(\hbar\omega)\, d(\hbar\omega) \tag{6.128}$$

is a measure of the width of the spectrum. For a crystal initially in thermal equilibrium, Av_α in the expression of $I_{ba}(\hbar\omega)$ is statistical averaging according to the temperature. Therefore, both the "frequency shift" $\langle \hbar\omega \rangle - \hbar\omega_{ba}$ and the width of the spectrum $\langle (\hbar\omega - \langle \hbar\omega \rangle)^2 \rangle$ depend on the temperature. It has been shown theoretically that the frequency shift would be independent of temperature if ω_j^a were equal to ω_j^b.

The transitions occurring at $\hbar\omega \simeq \hbar\omega_{ba}$ correspond to

$$\sum_{j=1}^N [(n_j^b + \tfrac{1}{2})\hbar\omega_j^b - (n_j^a + \tfrac{1}{2})\hbar\omega_j^a] = 0. \tag{6.129}$$

The condition signifies $n_j^b = n_j^a$ in the approximation of $\omega_j^b = \omega_j^a$. Transitions that are not associated with a change in n_j may be called zero-phonon transitions. They produce a peak in the spectrum whose prominence is stronger for weaker electron–phonon coupling.

The spectrum of transitions has an interesting structure if there are vibration modes that have strong interaction with the electron and have approximately the same frequency. Such is the case of an ionic crystal, the polar optical modes of which interact electrostatically with the electron by virtue of their polarization and have approximately the same frequency ω_0. As a result, the transition spectrum may exhibit a series of sharp peaks that differ successively by $\hbar\omega_0$ in energy. The problem was treated first in connection with the optical absorption of F centers in alkali halides.

An often-used approximation is the Franck–Condon principle, which assumes that the ions stand still while the electron makes the transition. In this approximation, we have

$$I_{ba}(\hbar\omega) = |(H_{\mathrm{I}})_{ba}|^2 \mathrm{Av}_\alpha \int d\mathbf{R} |\chi_{a\alpha}(\mathbf{R})|^2 \delta[E_b(\mathbf{R}) - E_{a\alpha} \mp \hbar\omega]. \quad (6.130)$$

Furthermore, the energies $E_a(\mathbf{R})$ and $E_b(\mathbf{R})$ depend mainly on the configuration of the imperfection center and surrounding ions. Therefore they may be considered functions of "configurational coordinates" $\boldsymbol{\rho}$: $E_b(\boldsymbol{\rho})$ and $E_a(\boldsymbol{\rho})$. The wavefunction $\chi_a(\boldsymbol{\rho})$ to be considered pertains to the so-called local modes. It is usually sufficient to consider one or a few kinds of configurational coordinates. The Franck–Condon approximation model in conjunction with configurational coordinates simplifies the calculation and the understanding of the transitions.

e. Optical Effect of Lattice Vibration

In a crystal consisting of positively charged and negatively charged ions, the relative displacement of the two types of ions gives rise to electrical polarization. As mentioned in Section V.1.b, a displacement wave whose wavelength is large compared to the lattice constant may be considered in terms of local displacement, displacement in a unit cell. Let the local displacement of vibration mode $\mathbf{q}p$ be denoted by $\mathbf{w}_{\mathbf{q}p} M_{\mathbf{q}p}^{-1/2}$. The polarization and electric field are involved in the dynamics of ion vibration. They were not taken into account in Section II.2. In order to include their effects, consider two equations for a mode of vibration:

$$\ddot{\mathbf{w}} = -\omega^2 \mathbf{w} = b_{11}\mathbf{w} + b_{12}\mathbf{E}, \quad (6.131)$$

$$\mathbf{P} \equiv \frac{\varepsilon - 1}{4\pi}\mathbf{E} = b_{21}\mathbf{w} + b_{22}\mathbf{E}. \quad (6.132)$$

$\mathbf{P}$ is the polarization, ω is the frequency of vibration, ε is the dielectric constant, and the tensors $b_{11}, b_{12}, b_{21}, b_{22}$ are coefficients. Solving the two equations for $\mathbf{E}$, we get

$$\mathbf{P} = \frac{\varepsilon - 1}{4\pi}\mathbf{E} = \left(\frac{b_{21}b_{12}}{-\omega^2 - b_{11}} + b_{22}\right)\mathbf{E}. \tag{6.133}$$

Introduce the notations:

$$\varepsilon_\infty = 1 + 4\pi b_{22}, \qquad \varepsilon_0 = \varepsilon_\infty - 4\pi\frac{b_{21}b_{12}}{b_{11}}, \qquad \omega_0^2 = -b_{11}. \tag{6.134}$$

The expression

$$\varepsilon = \varepsilon_\infty + \frac{\varepsilon_0 - \varepsilon_\infty}{1 - (\omega/\omega_0)^2} \tag{6.135}$$

shows the frequency dispersion of ε. In particular,

$$\varepsilon(\omega = 0) = \varepsilon_0 \qquad \text{and} \qquad \varepsilon(\omega = \infty) = \varepsilon_\infty. \tag{6.136}$$

The coefficients $b_{11}, b_{12}, b_{21}, b_{22}$ and the parameters $\varepsilon_0, \varepsilon_\infty, \omega_0^2$ are in general tensor quantities. They may reduce to scalars, depending on the symmetry of the crystal structure.

An oscillation of electrical polarization is associated with an electromagnetic field. In the equation of motion, Eq. (6.131), the effect of the magnetic field is, as usual, considered negligible. The electric field $\mathbf{E}$, however, obeys Maxwell's field equations. Consider plane-polarized plane waves of $\mathbf{w}$, $\mathbf{P}$, and $\mathbf{E}$ with wavevectors $\mathbf{q}$ and frequencies ω. The field equations take the following form:

$$\mathbf{q}\cdot(\mathbf{E} + 4\pi\mathbf{P}) = 0, \tag{6.137a}$$

$$\mathbf{q}\cdot\mathbf{H} = 0, \tag{6.137b}$$

$$\mathbf{q}\times\mathbf{E} = \frac{\omega}{c}\mathbf{H}, \tag{6.137c}$$

$$\mathbf{q}\times\mathbf{H} = -\frac{\omega}{c}(\mathbf{E} + 4\pi\mathbf{P}). \tag{6.137d}$$

For such waves,

$$\mathbf{q}\perp\mathbf{H}\perp\mathbf{E}, \qquad \mathbf{q}\perp(\mathbf{E} + 4\pi\mathbf{P})\perp\mathbf{H}, \tag{6.138}$$

and the four equations are not independent, since the first follows from the last and the second follows from the third. Substituting expression (6.132) for $\mathbf{P}$ into the last equation, we get

$$\mathbf{q}\times\mathbf{H} = -\frac{\omega}{c}\left(1 + 4\pi b_{22} + \frac{4\pi b_{21}b_{12}}{-b_{11} - \omega^2}\right)\mathbf{E}. \tag{6.139}$$

On the other hand, the cross product of the third equation with $\mathbf{q}$ gives

$$\frac{\omega}{c}\mathbf{q}\times\mathbf{H}=\mathbf{q}\times(\mathbf{q}\times\mathbf{E})=\mathbf{q}(\mathbf{q}\cdot\mathbf{E})-\mathbf{E}q^2. \tag{6.140}$$

It follows that

$$\left(\frac{qc}{\omega}\right)^2\left[\mathbf{E}-\mathbf{q}\left(\frac{\mathbf{q}\cdot\mathbf{E}}{q^2}\right)\right]=\left(1+4\pi b_{22}+\frac{4\pi b_{21}b_{12}}{-b_{11}-\omega^2}\right)\mathbf{E}. \tag{6.141}$$

The equation gives the interrelationship of $\mathbf{q}$, $\mathbf{E}$, and ω.

Equations (6.137c) and (6.137d) show that plane-polarized plane waves of $\mathbf{q}=0$ can always have $\omega=0$. A finite frequency is possible under the condition

$$0=\mathbf{E}+4\pi\mathbf{P}=\left(1+4\pi b_{22}+\frac{4\pi b_{21}b_{12}}{-b_{11}-\omega^2}\right)\mathbf{E}. \tag{6.142}$$

The condition

$$1+4\pi b_{22}+\frac{4\pi b_{21}b_{12}}{-b_{11}-\omega^2}=0 \tag{6.143}$$

leads to

$$\omega^2=-b_{11}+\frac{4\pi b_{21}b_{12}}{1+4\pi b_{22}}=\left(\frac{\varepsilon_0}{\varepsilon_\infty}\right)\omega_0^2\equiv\omega_{\mathrm{L}}^2. \tag{6.144}$$

This expression of ω^2 must naturally be scalar. It is obviously scalar if the coefficients b are scalars, which is the case for many crystals of common interest. For simplicity, we shall henceforth assume ω_{L} to be scalar.

The wave functions (6.137) admit solutions with

$$\mathbf{E}+4\pi\mathbf{P}=0, \tag{6.145}$$

regardless of $\mathbf{q}$. The frequency $\omega=\omega_{\mathrm{L}}$ of such waves is independent of $\mathbf{q}$. Equation (6.137d) becomes $\mathbf{q}\times\mathbf{H}=0$, which in conjunction with (6.137b) leads to

$$\mathbf{H}=0. \tag{6.146}$$

It follows then from (6.137c) and (6.137a) that

$$\mathbf{q}\parallel\mathbf{E}\parallel\mathbf{P}. \tag{6.147}$$

The field is characterized only by a scalar potential with a longitudinal $\mathbf{E}$. This is the case considered for the electron scattering by polar modes of lattice vibration in Section V.1.b. The effective charge e^* introduced in (5.17) represents the relationship between $\mathbf{E}$ and the relative displacement $\mathbf{w}M^{-1/2}$ of oppositely charged ions:

$$\mathbf{E}_{\mathbf{q}p} = -(4\pi e^* \mathbf{w} M^{-1/2})_{\mathbf{q}p}. \tag{6.148}$$

We get the following expression for e^*:

$$e^* = M_{\mathbf{q}p}^{1/2}\left(\frac{b_{21}}{1+4\pi b_{22}}\right)_{\mathbf{q}p} = M_{\mathbf{q}p}^{1/2}\left\{\omega_{\mathrm{L}}\left[\frac{1}{4\pi}\left(\frac{1}{\varepsilon_\infty}-\frac{1}{\varepsilon_0}\right)\right]^{1/2}\right\}_{\mathbf{q}p}. \tag{6.149}$$

We note that e^* may not be a scalar merely because ω_{L} is assumed to be scalar. If all the coefficients b are scalars, e^* must be scalar, and the ion displacement $\mathbf{w}$ is longitudinal: $\mathbf{w} \parallel \mathbf{E} \parallel \mathbf{P} \parallel \mathbf{q}$.

Since Eqs. (6.137) are linear in the field vectors, we may divide the waves into two categories. The category of $\mathbf{E} \parallel \mathbf{q}$ has been discussed in the preceding paragraph. Consider now waves with $\mathbf{E} \perp \mathbf{q}$. With $\mathbf{q} \cdot \mathbf{E} = 0$, (6.141) gives

$$\left(\frac{qc}{\omega}\right)^2 = 1 + 4\pi b_{22} + \frac{4\pi b_{21} b_{12}}{-b_{11} - \omega^2} = \varepsilon_\infty \frac{\omega^2 - \omega_{\mathrm{L}}^2}{\omega^2 - \omega_0^2}. \tag{6.150}$$

For a given q, the quadratic equation of ω gives two solutions of frequency. Denote the larger and smaller frequencies by ω_+ and ω_-, respectively. For large wavevectors, $q \gg (\varepsilon_0^{1/2}/c)\omega_0$, we have

$$\omega_+ \simeq cq/\varepsilon_\infty^{1/2} \qquad \text{and} \qquad \omega_- \simeq \omega_0. \tag{6.151}$$

As q approaches zero, ω_+ approaches ω_{L} and ω_- tends to 0. The field with $\mathbf{E} \perp \mathbf{q}$ and consequently $\mathbf{H} \neq 0$ is an electromagnetic field with a scalar and a vector potential. The normal modes of the system, lattice vibration with electromagnetic field, are called polaritons. $\omega_+(\mathbf{q})$ and $\omega_-(\mathbf{q})$ are frequencies of the two branches of polaritons.

The refractive index n and the extinction coefficient k of the electromagnetic waves are related to the dielectric function by

$$n^2 - k^2 = \varepsilon_1 \qquad \text{and} \qquad 2nk = \varepsilon_2. \tag{6.152}$$

The preceding discussion leads to $\varepsilon_1 = \varepsilon$ and $\varepsilon_2 = 0$, with $\varepsilon(\omega)$ given by (6.135). Beginning with ε_0, the dielectric function $\varepsilon(\omega)$ increases slowly with increasing ω in the range $0 < \omega < \omega_0$, then rapidly toward $+\infty$ at $\omega = \omega_0$. As ω increases beyond the resonant frequency ω_0, $\varepsilon(\omega)$ increases from $-\infty$ at $\omega = \omega_0$, passes through zero at $\omega = \omega_{\mathrm{L}}$, and approaches a constant value ε_∞ at high ω. Waves with frequencies in the range $\omega_0 < \omega < \omega_{\mathrm{L}}$ do not exist. In this frequency range, $\varepsilon < 0$, and consequently $n = 0$ and $k = (-\varepsilon)^{1/2}$. The crystal has a reflecting power $R = 1$; incident electromagnetic waves with frequencies in this range are totally reflected. Waves with $\omega < \omega_0$ belong to the lower branch of polaritons; the index of refraction for their propagation, $n = \varepsilon^{1/2}$, approaches $\varepsilon_0^{1/2}$ as ω tends to zero. Waves with $\omega > \omega_L$ belong to the upper branch of polaritons; their refractive index approaches $\varepsilon_\infty^{1/2}$ at high frequencies.

Anharmonicity of Lattice Vibration. Equation (6.131) is based on the harmonic vibration of the lattice, an approximation that makes it possible to consider each mode of vibration separately. Actually, phonons of different modes are coupled through anharmonicity. In a first approximation, this complication can be taken into account by adding a term

$$\Gamma \dot{\mathbf{w}} = -i\omega\Gamma\mathbf{w}$$

to the left-hand side of (6.131), the equation for a single mode. The constant Γ may be referred to as the damping frequency. Expression (6.135) for the dielectric function now becomes

$$\varepsilon = \varepsilon_\infty + \frac{(\varepsilon_0 - \varepsilon_\infty)\omega_0^2}{\omega_0^2 - (\omega^2 + i\Gamma\omega)},$$

with

$$\varepsilon_1 = \varepsilon_\infty + \frac{(\varepsilon_0 - \varepsilon_\infty)(\omega_0^2 - \omega^2)\omega_0^2}{(\omega_0^2 - \omega^2)^2 + \Gamma^2\omega^2} \tag{6.153}$$

and

$$\varepsilon_2 = \frac{(\varepsilon_0 - \varepsilon_\infty)\omega_0^2\omega\Gamma}{(\omega_0^2 - \omega^2)^2 + \Gamma^2\omega^2}. \tag{6.154}$$

In addition to modifying ε_1, the presence of Γ gives rise to ε_2, which represents absorption of the electromagnetic wave. ε_2 decreases toward zero as ω approaches zero, and it becomes $(\varepsilon_0 - \varepsilon_\infty)/\omega_0^2\Gamma/\omega^3$ for $\omega \gg (\omega_0 + \Gamma)$. When $\Gamma \ll \omega_0$, the peak of absorption occurs at a frequency close to ω_0.

3. INTERBAND TRANSITIONS OF ELECTRONS

As shown in (6.79) and (6.82), the excitation of an electron from state $|m\mathbf{k}\rangle$ to an unoccupied state $|m'\mathbf{k}'\rangle$ by radiation of frequency ω is subject to the conditions

$$\mathbf{k}' = \mathbf{k} + \boldsymbol{\kappa} \approx \mathbf{k} \tag{6.155}$$

and

$$\hbar\omega = \epsilon(m'\mathbf{k}') - \epsilon(m\mathbf{k}) \equiv \hbar\omega_{\mathbf{k},m'm}. \tag{6.156}$$

The minimum of $\epsilon(m'\mathbf{k}) - \epsilon(m\mathbf{k})$ gives the threshold, the beginning, of electron transitions from band m to m'. The contribution of the transitions to the optical parameters provides information on the structures of the two energy bands, as can be seen in (6.90) and (6.91). The contribution to ε_2, in particular, can be written as

$$\epsilon_2 = 2\left\{4\pi\frac{\pi e^2}{2m\omega}\frac{1}{(2\pi)^3}\int_S ds[f(\epsilon_{m\mathbf{k}}) - f(\epsilon_{m'\mathbf{k}})]\tilde{f}_{\mathbf{k},m'm}\left(\frac{\partial\omega_{\mathbf{k},m'm}}{\partial k_\perp}\right)^{-1}\right\}, \tag{6.157}$$

where the factor 2 in front of the curly brackets accounts for the two spins of an electron, S is the surface of $\omega_{\mathbf{k},m'm} = \omega$ in $\mathbf{k}$ space, and $k_\perp$ is the normal of ds. With increasing frequency, transitions of various pairs of energy bands will be superimposed successively.

The interband transitions that have the lowest threshold frequency stand out clearly up to the next threshold of other interband transitions. For a metal, such transitions are those either from or to the conduction band, giving useful information about the Fermi surface. However, even at frequencies below the second threshold of interband transitions, the situation is complicated by the superposition of the free-carrier effect of conduction electrons, an effect that can be quite strong. In an insulator or a semiconductor, transitions from the valence band to the conduction band have the lowest threshold frequency. Since the effect of free carriers is low owing to the sparsity of carriers, the effect of the interband transitions predominates practically down to the threshold frequency. The steep rise of optical absorption, which occurs as the frequency is increased beyond the threshold, is called the fundamental absorption edge. Studies of the absorption edge provide information on the processes associated with the energy gap between the valence band and the conduction band. Consider, for example, the following simple case: The minimum energy of the conduction band and the maximum energy of the valence band occur at the same wavenumber $\mathbf{k}_0$, and the energy bands are approximately isotropic with scalar effective masses m_c and m_v. We then have

$$\begin{aligned}\hbar\omega_{\mathbf{k},cv} &= \epsilon_g + \frac{\hbar^2}{2}\left(\frac{1}{m_c} + \frac{1}{m_v}\right)(\mathbf{k} - \mathbf{k}_0)^2 \\ &\equiv \epsilon_g + \frac{\hbar^2}{2m_r}(\mathbf{k} - \mathbf{k}_0)^2,\end{aligned} \tag{6.158}$$

ϵ_g being the energy gap. Surfaces of constant $\hbar\omega_{\mathbf{k},cv}$ are spherical surfaces centered at $\mathbf{k}_0$. The interband transitions give a scalar ε_2. With

$$f_{\mathbf{k},cv} = \frac{2}{m}\frac{|[(\boldsymbol{\kappa}/\kappa)\cdot\mathbf{p}_{\mathbf{k},cv}]|^2}{\hbar\omega_{\mathbf{k},cv}},$$

we get from (6.157)

$$\varepsilon_2 = \frac{2e^2(2m_r)^{3/2}}{3m^2\hbar}|\mathbf{p}_{cv}(\hbar\omega)|^2(\hbar\omega)^{-2}(\hbar\omega-\epsilon_g)^{1/2}. \tag{6.159}$$

Effect of excitons. A photon $\hbar\omega$ too small for producing an electron and a separate hole as in an interband transition may be adequate for the excitation of an electron–hole pair, an exciton. An absorption edge begins with the production of excitons rather than an electron and a hole unbound to each other. In some ionic insulators especially, the exciton effect extends over a considerable range of ω, showing interesting structures.

Indirect transitions. As in the case of the intraband effect of free carriers, the perturbation H_{I} of radiation may, in combination with some inherent interaction H_0' of the crystal, produce interband transitions of electrons in second-order perturbation. Interband transitions of this kind have come to be called indirect transitions, in contrast to direct transitions due to H_{I} alone. The weak effect of indirect transitions gains significance if the minimum energy and the maximum energy, respectively, of the two bands do not occur at the same wavenumber $\mathbf{k}$. In this case, direct transitions begin at $\hbar\omega > \epsilon_g$, owing to condition (6.155), whereas indirect transitions may begin at a $\hbar\omega$ at or close to ϵ_g. The rate of an indirect transition is given by (6.95). Instead of (6.155) and (6.156), the conditions required for an indirect transition are

$$\mathbf{k}_{\mathrm{F}} = \mathbf{k}_0 + \Delta\mathbf{k} \tag{6.160}$$

and

$$\hbar\omega = \epsilon_{\mathrm{F}} - \epsilon_0 + \Delta E. \tag{6.161}$$

$\Delta\mathbf{k}$ and ΔE are, respectively, the change in the electron wavenumber associated with H_0' and the change in energy due to the involvement of H_0'. Consider the case where H_0' is the interaction of the electron with lattice vibration. We then have

$$\Delta\mathbf{k} = \mp\mathbf{q} \qquad \text{and} \qquad \Delta E = \pm\hbar\omega_{\mathbf{q}}, \tag{6.162}$$

where the upper and lower signs apply to the emission and absorption, respectively, of a phonon of wavevector $\mathbf{q}$ and energy $\hbar\omega_{\mathbf{q}}$. Let $\mathbf{k}_{hl}$ be the wavevector difference between the minimum of the higher energy band and the maximum of the lower energy band. Clearly, interband transitions for the two bands begin at

$$\hbar\omega = \epsilon_g \pm \hbar\omega_{\mathbf{q}=\mathbf{k}_{hl}}. \tag{6.163}$$

Various phonon branches with different $\omega(\mathbf{q})$ give different thresholds for indirect transitions, leading to a ladder structure that adjoins the steeply rising direct transitions at a higher radiation frequency ω. Such a phenomenon has been observed in studies of the fundamental absorption edge in some semiconductors.

4. NONLINEAR OPTICAL PROPERTIES

In response to the electric field of radiation,

$$\mathbf{E}(\boldsymbol{\kappa},\omega) + \mathbf{E}^*(\boldsymbol{\kappa},\omega) = \mathcal{E}\exp[i(\boldsymbol{\kappa}\cdot\mathbf{r} - \omega t)] + \mathcal{E}^*\exp[-i(\boldsymbol{\kappa}\cdot\mathbf{r} - \omega t)], \tag{6.164}$$

an electric current, or alternatively an electric polarization $\mathbf{P} + \mathbf{P}^*$, is produced in the material, which has components of various frequencies and wavevectors:

$$\mathbf{P} = \tilde{\chi}(\omega)\cdot\mathbf{E} + \tilde{\chi}(0 = \omega - \omega) : \mathbf{E}\mathbf{E}^* + \tilde{\chi}(2\omega = \omega + \omega) : \mathbf{E}\mathbf{E}$$
$$+ \tilde{\chi}(\omega = \omega + \omega - \omega) \vdots \mathbf{E}\mathbf{E}\mathbf{E}^* + \tilde{\chi}(3\omega = \omega + \omega + \omega) \vdots \mathbf{E}\mathbf{E}\mathbf{E} + \cdots. \tag{6.165}$$

All the $\tilde{\chi}$'s except the one of the first term are nonlinear susceptibilities for correspondingly different harmonic frequencies. They represent the radiation–material interaction to higher than the first order of perturbation. A component of $\mathbf{P}$ that involves a larger number of $\mathbf{E}$'s becomes significant at larger values of E. There is a field $\mathbf{E}(\omega_s)$ associated with a polarization $\mathbf{P}(\omega_s)$ according to the equation

$$\nabla\times\nabla\times\mathbf{E}(\omega_s) - \frac{\tilde{\varepsilon}(\omega_s)\omega_s^2}{c^2}\mathbf{E}(\omega_s) = \frac{4\pi\omega_s^2}{c^2}\mathbf{P}(\omega_s). \tag{6.166}$$

Thus we see that a material generates harmonic waves of radiation by virtue of its nonlinear susceptibilities.

Under the application of two plane waves of radiation $\mathbf{E}(\boldsymbol{\kappa}_1,\omega_1)$ and $\mathbf{E}(\boldsymbol{\kappa}_2,\omega_2)$, harmonic polarizations of each frequency, ω_1 and

ω_2, are produced. In addition, there will be polarizations of various combination frequencies, $\omega_1 - \omega_2$, $\omega_1 + \omega_2$, $2\omega_1 - \omega_2$, $2\omega_2 - \omega_1$, $2\omega_1 + \omega_2$, and so on, that generate radiation waves of combination frequencies.

The calculation of linear susceptibility is covered in the derivation of (6.83) for the dielectric function $\tilde{\varepsilon}$. The nonlinear susceptibilities can be calculated analogously by extending the calculations to higher orders of radiation field perturbation. Expressions can be found in the literature; their complexity increases rapidly with increasing rank of the susceptibility tensor. The following discussion is limited to the consideration of transition processes involved in the nonlinear properties. Take, for example, the generation of a wave of combination frequency $\omega_1 + \omega_2$, resulting from electron excitations. The one-electron states involved are the initial state $|0\rangle$, two intermediate states $|i'\rangle$ and $|i''\rangle$, and the final state $|f\rangle$. The $|0\rangle$ to $|i'\rangle$ and $|i'\rangle$ to $|i''\rangle$ transitions correspond, respectively, to the absorption of a $\hbar\omega_1$ and a $\hbar\omega_2$. The $|i''\rangle$ and $|f\rangle$ corresponds to the emission of a $\hbar\omega_1 + \hbar\omega_2$ photon. According to the condition of energy conservation, the one-electron energies satisfy the relation

$$\epsilon_f = \epsilon_0. \tag{6.167}$$

The wavevector conservation of the transition matrix elements leads to

$$\mathbf{k}_f = \mathbf{k}_0 + \boldsymbol{\kappa}_1 + \boldsymbol{\kappa}_2 - \boldsymbol{\kappa}_{(1+2)}. \tag{6.168}$$

The rate of transition from $|0\rangle$ to $|f\rangle$, which leads to the generation of radiation with wavevector $\boldsymbol{\kappa}_{(1+2)}$ and frequency $\omega_1 + \omega_2$, can be calculated from third-order perturbation theory. The calculation should include multiplication by

$$f_0(1 - f_{i'})(1 - f_{i''})(1 - f_f) \tag{6.169}$$

for statistical consideration; for $\boldsymbol{\kappa}_{(1+2)} = \boldsymbol{\kappa}_1 + \boldsymbol{\kappa}_2$, $\mathbf{k}_f = \mathbf{k}_0$, $|f\rangle = |0\rangle$, and the factor $1 - f_f$ should be omitted. Finally, the result so obtained must be summed over the initial one-electron states $|0\rangle$.

D. MAGNETIC PROPERTIES

1. GENERAL CONSIDERATIONS

The magnetic state of a substance is characterized by the magnetic moment per unit volume, or magnetization, $\mathbf{M}$. The response of matter to an applied magnetic field **H** is represented by the magnetic susceptibility $\tilde{\chi}$:

$$\mathbf{M} = \tilde{\chi} \cdot \mathbf{H}. \tag{6.170}$$

An increment $\Delta\mathbf{H}$ of the applied field changes the energy of the matter (per unit volume) by

$$\Delta E = -\mathbf{M} \cdot \Delta\mathbf{H} = -(\tilde{\chi} \cdot \mathbf{H}) \cdot \Delta\mathbf{H}.$$

For $\mathbf{H} \| \Delta\mathbf{H} \| \hat{h}$, we get

$$\tilde{\chi} \cdot \hat{h} = -\frac{1}{H}\frac{\partial E}{\partial \mathbf{H}}. \tag{6.171}$$

A system can be in equilibrium under an applied magnetic field. For the canonical ensemble of a system in equilibrium, the density matrix has the form

$$\rho_{ii'} = \exp(-E_i/kT)\delta_{ii'},$$

and the statistical value of the susceptibility is given by

$$\tilde{\chi} \cdot \hat{h} = \sum_i \left(-\frac{1}{H}\frac{\partial F_i}{\partial \mathbf{H}}\right) \frac{\exp(-E_i/kT)}{\sum \exp(-E_i/kT)} = \frac{kT}{H}\frac{\partial \ln f}{\partial \mathbf{H}} = -\frac{1}{H}\frac{\partial F}{\partial \mathbf{H}}, \tag{6.172}$$

where

$$f = \sum_i \exp(-E_i/kT)$$

is the partition function and F is the Helmholtz free energy for a unit volume of the system.

A susceptibility, or one of its components, that is positive in sign indicates that the induced magnetization is parallel to the applied magnetic field; it is referred to as paramagnetic. A susceptibility, or its component, that is negative in sign is called diamagnetic. A material may be in a state that shows spontaneous magnetization. There are three types of spontaneous magnetization: ferromagnetism, antiferromagnetism, and ferrimagnetism. In ferromagnetism, all the pertinent constituents of the material have a component of magnetic moment in the same direction. In antiferromagnetism, two groups of constituents have magnetizations that are equal in magnitude but opposite in direction. In ferrimagnetism, two oppositely oriented magnetizations are different in magnitude, giving a net magnetization for the material.

2. MAGNETIC SUSCEPTIBILITY OF FREE CARRIERS

According to classical mechanics, charged particles that are contained in a fixed volume but are otherwise free have zero magnetic susceptibility. This conclusion follows from the consideration that a magnetic field affects the velocity but not the speed or energy of the particles. Consequently, the magnetic susceptibility is zero according to (6.171). For particles that each have an intrinsic magnetic moment $\mathbf{m}$, such as electrons with $\mathbf{m} = 2\beta\mathbf{s}$, an applied magnetic field $\mathbf{B}$ affects the particle energy by $-\mathbf{m} \cdot \mathbf{B}$, making the particles prefer to have $\mathbf{m}$ parallel rather than antiparallel to $\mathbf{B}$. Quantitatively, the effect depends on the energy distribution of the particles. For free electrons at temperatures so low that the Fermi–Dirac distribution is near complete degeneracy, the paramagnetic susceptibility is given by

$$\chi = M/H = 2\beta^2 \rho(\epsilon_F) B/H, \tag{6.173}$$

where $\rho(\epsilon_F)$ is the density of states of electrons at the Fermi energy ϵ_F. This effect was first considered by Pauli for the conduction electrons in metals and is known as Pauli paramagnetism.

Quantum mechanically, the eigenstates of a free charged particle are affected by the introduction of a magnetic field, like the quantization of Landau levels discussed in Section IV.A.3. Therefore free carriers contribute to diamagnetic as well as paramagnetic susceptibility. Consider free carriers that are characterized by a scalar effective mass. This simple case is adequate to bring out in principle conclusions of significance. Under the assumption that the g factor is also a scalar, the spin term of the Hamiltonian given in (4.39) is $-\beta g \mathbf{s} \cdot \mathbf{B}$. We note that β for an electron with negative charge is negative in our notation. As pointed out before, the g factor may be quite large and may be negative for some materials. Owing to the spin term, each Landau subband is split into two bands. For the two bands split from a subband n, the one-electron energy may be written as

$$\epsilon = \frac{\hbar^2 k_z^2}{2m^*} + (n + \tfrac{1}{2})\hbar\omega_c \pm \frac{|g|\hbar\omega_0}{4}, \tag{6.174}$$

where $\omega_0 = |eB|/mc$ and the $\pm$ sign refers to the direction of spin with respect to direction $\hat{z}$ of $\mathbf{B}$. The density of states is given by (4.29) for each of the split subbands.

According to Fermi–Dirac statistics, the free energy of the system of carriers, per unit volume, is given by

$$F = \frac{N}{V}\epsilon_{\mathrm{F}} - kT\sum_i \ln\left[1+\exp\left(\frac{\epsilon_{\mathrm{F}}-\epsilon_i}{kT}\right)\right], \tag{6.175}$$

where N/V is the concentration of carriers. The running index i covers all the subbands, each with two branches split by spin. By using (6.174) and (4.29), it can be shown that the magnetization is given by

$$\begin{aligned} M &= \partial F/\partial H \\ &= M_0\left[\left(\frac{3g^2m^{*2}}{4m_0^2}-1\right)+\frac{6\pi kT}{\hbar\omega_{\mathrm{c}}}\left(\frac{2\epsilon_{\mathrm{F}}}{\hbar\omega_{\mathrm{c}}}\right)^{1/2}\right. \\ &\qquad \left.\times\sum_n \frac{(-1)^n}{n^{1/2}}\frac{\sin(2\pi n\epsilon_{\mathrm{F}}/\hbar\omega_{\mathrm{c}}-\pi/4)\cos(\pi ngm^*/2m_0)}{\sinh(2\pi^2nkT/\hbar\omega_{\mathrm{c}})}\right], \end{aligned} \tag{6.176}$$

where

$$M_0 = (e^2/12\pi^2\hbar c^2)(2\epsilon_{\mathrm{F}}/m^*)^{1/2}B.$$

The first term in the square brackets of M is independent of B. It contains a positive component and a negative component. The positive component corresponds to (6.173) of Pauli paramagnetism; it leads to (6.173) if the g factor is taken to be equal to 2. The negative component is a diamagnetism, which is absent without the Landau quantization of the energy levels. The second term in the square brackets of M oscillates with variation in B, as ϵ_{F} crosses the lowest energy $[(n+\frac{1}{2})\hbar\omega_{\mathrm{c}} \pm |g|\hbar\omega_0]$ of various spin-split subbands. The oscillation of magnetic susceptibility becomes clearly observable at low temperature. The phenomenon was first observed in bismuth, and it is called the de Haas–van Alphen effect. It provides an effective means for the study of Fermi surfaces.

It should be emphasized that the above discussion including Section IV.A.3 about Landau levels is based on the one-electron approximation. The magnetic field considered is the field **B** experienced by an individual electron in this approximation. The M and χ obtained depend on the acceptability of the conventional one-electron approximation in the absence of an externally applied magnetic field.

3. A SYSTEM OF ISOLATED ATOMS

Under an applied magnetic field $\mathbf{H}_a$, an atom may show a magnetic moment

$$\mathbf{M}_a = \chi_a \cdot \mathbf{H}_a, \tag{6.177}$$

where the atomic susceptibility χ_a is a scalar in simple cases. If Russell–Saunders coupling is applicable for the atom, then the statistical χ_a is given by

$$\chi_a = \frac{\beta g J}{H_a} \frac{(J+\frac{1}{2})\coth(J+\frac{1}{2})\alpha - \frac{1}{2}\coth(\alpha/2)}{J} = \frac{\beta g J}{H_a} B_J(\alpha), \tag{6.178}$$

in which $\beta = e\hbar/2mc$ is the Bohr magneton, g is the Landé g factor, J is the quantum number of total angular momentum, $\alpha = H_a \beta g/kT$, and $B_J(\alpha)$ is the Brillouin function.

Consider a system of atoms that are isolated in the sense that one atom is affected by the others only through the field given by the others at its location. The system is clear-cut in comparison with the case of conduction carriers in terms of one-electron approximation, an approximation that may be altered by an applied field. For a system of one kind of atoms arranged equivalently in a lattice, $\mathbf{H}_a$ and χ_a are each the same for all the atoms. The magnetic moment per unit volume of the system is

$$\mathbf{M} = n\chi_a \mathbf{H}_a, \tag{6.179}$$

n being the concentration of atoms. In order to obtain the susceptibility χ defined by $\mathbf{M} = \chi \cdot \mathbf{H}$, the problem is to determine the relationship between the so-called local field $\mathbf{H}_a$ and the macroscopic field $\mathbf{H}$.

The simplest relationship between $\mathbf{H}_a$ and $\mathbf{H}$ for such a system is that given by Lorentz:

$$\mathbf{H}_a = \mathbf{H} + \frac{4\pi}{3}\mathbf{M} = \mathbf{H} + \frac{4\pi}{3} n\chi_a \mathbf{H}_a. \tag{6.180}$$

The susceptibility of the system is then

$$\chi = n\chi_a \left[1 - \frac{4\pi}{3} n\chi_a\right]^{-1}. \tag{6.181}$$

Onsager modified Lorentz's treatment by taking into account the fact that the local field acting on an atom does not include the field of the atom itself. He considered a sphere centered at the atom under consideration, with a volume equal to the volume per atom, and the effect of the magnetic moment of the atom on those of the surrounding atoms, which depends on the positions of the surrounding atoms. The field produced by the medium in a hollow spherical cavity was taken to be $\mathbf{H}_a - \mathbf{H}$ for the atom in the cavity. The treatment gave

$$\mathbf{H}_a = \left(1 + \frac{4\pi\chi}{8\pi\chi + 3}\right)\mathbf{H}, \tag{6.182}$$

which leads to

$$\chi = \frac{3}{16\pi}\left[-1 + 4\pi n\chi_a + \left(1 + \frac{8\pi}{3}n\chi_a + 16\pi^2 n^2 \chi_a^2\right)^{1/2}\right]. \tag{6.183}$$

We will not go into the more refined treatments given subsequently by Van Vleck.

4. EFFECTS OF EXCHANGE INTERACTION

a. Heisenberg Exchange of Nearly Localized Electrons

A solid material is really not a collection of isolated atoms. The electrons, which may not be considered as localized on isolated ions, have exchange interactions among themselves; this has been discussed in connection with the Hartree–Fock approximation. Heisenberg realized that the exchange interaction may play an important role in the magnetic property of a material. According to Heisenberg, the Hamiltonian contains the term

$$H_{ex} = -\frac{2}{\hbar^2}\sum_{i>j} J_{ij}\mathbf{s}_i \cdot \mathbf{s}_j = -2\sum_{i>j} J_{ij}\mathbf{S}_i \cdot \mathbf{S}_j, \tag{6.184}$$

in which i and j are electron indices and $\mathbf{s} = \hbar\mathbf{S}$ is the operator of spin angular momentum. This term is called the Heisenberg exchange. Consider the case where the one-electron wavefunctions of relevant electrons are large only around the individual ions. The subscripts i and j may then be considered ion indices.

In order to elucidate the nature of this exchange, consider a system of two atoms, a and b, and two electrons, 1 and 2. The wavefunction of an individual atom consists of a function of coordinates

ψ and a spinor u. There are two types of antisymmetric wavefunctions for the system:

$$\Psi_{\mathrm{I}} = [\psi_{\mathrm{a}}(1)\psi_{\mathrm{b}}(2) + \psi_{\mathrm{a}}(2)\psi_{\mathrm{b}}(1)][u_+(1)u_-(2) - u_-(1)u_+(2)]$$

and

$$\Psi_{\mathrm{II}} = [\psi_{\mathrm{a}}(1)\psi_{\mathrm{b}}(2) - \psi_{\mathrm{a}}(2)\psi_{\mathrm{b}}(1)]\begin{cases} u_+(1)u_+(2) \\ [u_+(1)u_-(2) + u_-(1)u_+(2)] \\ u_-(1)u_-(2) \end{cases}. \tag{6.185}$$

The eigenvalue of $(\mathbf{S}_1 + \mathbf{S}_2)^2$ has the form $S(S+1)$ for each of the functions. The singlet state I has $S = 0$. Each state of triplet level II has $S = 1$. It can be shown straightforwardly that the energy eigenvalue of the system contains a term

$$-2J\langle \mathbf{S}_1 \cdot \mathbf{S}_2 \rangle = -\frac{1}{2\beta^2} J\langle \mathbf{m}_1 \cdot \mathbf{m}_2 \rangle, \tag{6.186}$$

where $\mathbf{m} = -2\beta\mathbf{S}$ is the operator of magnetic moment. The eigenvalue $\langle \mathbf{S}_1 \cdot \mathbf{S}_2 \rangle$ is $-\frac{3}{4}$ when $S = 0$, and it is $\frac{1}{4}$ for $S = 1$. The exchange integral J is given by the expression

$$J = \int \psi_{\mathrm{a}}^*(1)\psi_{\mathrm{b}}^*(2)\left(\frac{e^2}{r_{\mathrm{ab}}} + \frac{e^2}{r_{12}} - \frac{e^2}{r_{1\mathrm{b}}} - \frac{e^2}{r_{2\mathrm{a}}}\right)\psi_{\mathrm{a}}(2)\psi_{\mathrm{b}}(1)\,d\tau_{12}. \tag{6.187}$$

An energy difference between the singlet and triplet, $E_{\mathrm{I}} - E_{\mathrm{II}} = 2J$, is given by (6.186), which represents the Heisenberg exchange energy. Expression (6.186) appears to be given by the effect of the relative orientation of two magnetic moments. It actually comes from the electrostatic interaction of the electrons, the group of which has an antisymmetric wavefunction. This simple example serves to indicate that electrons with a positive J favor parallel alignment of their magnetic moments, which lowers the energy, whereas electrons with a negative J favor antiparallel alignment.

The effect of the Heisenberg exchange in a crystalline solid can be demonstrated by considering an idealized crystal in which the exchange interaction has the same expression for each of the ions. Furthermore, for simplicity we will focus on the interaction of an ion with its nearest neighbors and let the exchange interaction J be the same for each nearest neighbor. We then have

$$H_{\mathrm{ex}} = -J\sum_{i,\mathbf{r}} \mathbf{S}_i \cdot \mathbf{S}_{i+\mathbf{r}}, \tag{6.188}$$

where the vectors $\mathbf{r}$ connect ion i with its nearest neighbors.

b. Spin Waves

Ferromagnetic Interaction. In this case, J is positive. The wavefunction Ψ is a determinant of functions $\psi_i(\mathbf{r}, \mathbf{s}) = \varphi_i(\mathbf{r})u_i(\mathbf{s})$, i being the index of atoms. The ground state, the state of lowest energy for a given $\varphi(\mathbf{r})$, has all the spins $\mathbf{s}_i$ parallel to each other. Let E_0 and Ψ_0 denote, respectively, the energy and wavefunction of this ground state. Bloch introduced spin waves of excitation:

$$\Psi_{\mathbf{k}} = a_{\mathbf{k}} \sum_n \Psi_n \exp(i\mathbf{k} \cdot \mathbf{r}_n), \tag{6.189}$$

where Ψ_n differs from Ψ_0 only by the reversal of the spin associated with atom n, and $\mathbf{k}$ is the wavevector. We show in the following that the spin waves are eigenstates of the system, much like the normal modes of lattice vibration.

Two operators a_i^+, a_i defined by

$$\begin{aligned} S_{ix} + S_{iy} &= (2S)^{1/2}(1 - a_i^+ a_i/2S)^{1/2} a_i, \\ S_{ix} - S_{iy} &= (2S)^{1/2} a_i^+ (1 - a_i^+ a_i/2S)^{1/2} \end{aligned} \tag{6.190}$$

have the commutation relation

$$[a_i, a_j^+] = \delta_{ij}. \tag{6.191}$$

The quantity S in the above expressions is given by $\mathbf{S}_i \cdot \mathbf{S}_i = S(S+1)$; $S = \frac{1}{2}$ in our case of one electron per atom. Introduce operators $b_{\mathbf{k}}^+, b_{\mathbf{k}}$ for a wavevector $\mathbf{k}$:

$$b_{\mathbf{k}} = N^{-1/2} \sum_i \exp(i\mathbf{k} \cdot \mathbf{R}_i) a_i, \qquad b_{\mathbf{k}}^+ = N^{-1/2} \sum_i \exp(-i\mathbf{k} \cdot \mathbf{R}_i) a_i^+, \tag{6.192}$$

where $\mathbf{R}_i$ is the position vector of atom i and N is the number of atoms. These operators have the following commutation relations:

$$[b_{\mathbf{k}}, b_{\mathbf{k}'}^+] = \delta_{\mathbf{k}\mathbf{k}'}, \qquad [b_{\mathbf{k}}, b_{\mathbf{k}'}] = [b_{\mathbf{k}}^+, b_{\mathbf{k}'}^+] = 0. \tag{6.193}$$

The Heisenberg exchange expressed in terms of $b_{\mathbf{k}}^+$ and $b_{\mathbf{k}}$ is

$$H_{\text{ex}} = -JNnS^2 + \sum_{\mathbf{k}} 2JnS\left(1 - n^{-1} \sum_{\mathbf{r}} \exp(i\mathbf{k} \cdot \mathbf{r})\right) b_{\mathbf{k}}^+ b_{\mathbf{k}} + H_1, \tag{6.194}$$

where n is the number of nearest neighbors of an atom and $\mathbf{r}$ is the position vector from the atom to a neighbor. H_1 consists of terms of higher order in $b_{\mathbf{k}}^+, b_{\mathbf{k}}$, and it may be neglected when the number of spin waves excited is not excessive.

It is of interest to consider the problem in the presence of a static magnetic field $\mathbf{B}_0$. The results obtained are then more general, covering the case of an applied magnetic field. The spatial orientation of the coordinate system is arbitrary in our simple model. With a $\mathbf{B}_0$, its direction can be taken as the z axis. In terms of the operators introduced in (6.190) and (6.192),

$$S_{iz} = S - a_i^+ a_i = S - N^{-1} \sum_{\mathbf{k}\mathbf{k}'} \exp[i(\mathbf{k} - \mathbf{k}') \cdot \mathbf{R}_i] b_{\mathbf{k}}^+ b_{\mathbf{k}'}. \tag{6.195}$$

The operator for the z component of total spin of the ions is

$$S_z = NS - \sum_{\mathbf{k}} b_{\mathbf{k}}^+ b_{\mathbf{k}}. \tag{6.196}$$

In the presence of $\mathbf{B}_0$, the Hamiltonian to be considered is

$$H_{\text{ex}} - \sum_i g\beta B_0 S_{zi} = H_{\text{ex}} - g\beta B_0 NS + g\beta B_0 \sum_{\mathbf{k}} b_{\mathbf{k}}^+ b_{\mathbf{k}}. \tag{6.197}$$

It can be shown that a linear combination of spin waves $\Psi_{\mathbf{k}}$ is an eigenfunction of this Hamiltonian. The energy eigenvalue E is given by

$$E - (-JNnS^2 - g\beta B_0 NS) = E - E_0 = \sum_{\mathbf{k}} n_{\mathbf{k}} \hbar\omega_{\mathbf{k}}. \tag{6.198}$$

The number $n_{\mathbf{k}}$ is a positive integer; it is the eigenvalue of $b_{\mathbf{k}}^+ b_{\mathbf{k}}$.

$$\hbar\omega_{\mathbf{k}} = \hbar \left[2JnS \left(1 - n^{-1} \sum_{\mathbf{r}} \exp(i\mathbf{k} \cdot \mathbf{r}) \right) + g\beta B_0 \right] \tag{6.199}$$

is the energy quantum of spin wave $\Psi_{\mathbf{k}}$; it is called a magnon. The operators $b_{\mathbf{k}}^+$ and $b_{\mathbf{k}}$ are the creation and annihilation operators, respectively, of magnon $\mathbf{k}$. The z component of total spin, S_z, has the eigenvalue $NS - \sum_{\mathbf{k}} n_{\mathbf{k}}$. In the ground state, $n_{\mathbf{k}} = 0$, the spins are perfectly aligned.

Antiferromagnetic Interaction. Consider now the effect of H_{ex} given by (6.188) when J is negative. In this case, the spin structure of the crystal is divided into two sublattices. All the nearest neighbors of an ion on one sublattice belong to the other sublattice. Introduce a fictitious magnetic field $\mathbf{H}_A$ to approximate the effect of crystal anisotropy. The anisotropy prescribes the direction along which spins of the two sublattices tend to align oppositely. Take $\hat{z} \parallel \mathbf{H}_\mathrm{A}$. The Hamiltonian to be considered is

$$H = -J\sum_{i,\mathbf{r}} \mathbf{S}_i \cdot \mathbf{S}_{i+\mathbf{r}} - g\beta H_\mathrm{A}\sum_i S_{iz}^a + g\beta H_\mathrm{A}\sum_i S_{iz}^b, \qquad (6.200)$$

where the superscript a or b specifies the sublattice.

Let N be the number of atoms in each sublattice. Introduce a set of operators $c_\mathbf{k}^+, c_\mathbf{k}$ for sublattice a, just as the set of operators $b_\mathbf{k}^+, b_\mathbf{k}$ were introduced in the previous case. For sublattice b, introduce operators $d_\mathbf{k}^+, d_\mathbf{k}$, which are defined by

$$d_\mathbf{k} = N^{-1/2}\sum_i \exp(-i\mathbf{k}\cdot\mathbf{R}_i) b_i$$

and

$$d_\mathbf{k}^+ = N^{-1/2}\sum_i \exp(i\mathbf{k}\cdot\mathbf{R}_i) b_i^+, \qquad (6.201)$$

in which the operators b_i^+, b_i are in turn defined by

$$S_{ix}^b + iS_{iy}^b = (2S)^{1/2} b_i^+ (1 - b_i^+ b_i/2S)^{1/2}$$

and

$$S_{ix}^b - iS_{iy}^b = (2S)^{1/2}(1 - b_i^+ b_i/2S)^{1/2} b_i. \qquad (6.202)$$

The commutation relations of operators $c_\mathbf{k}$ and operators $d_\mathbf{k}$ are similar to (6.193). In terms of the introduced operators, the Hamiltonian becomes

$$H = 2NnJS^2 - 2Ng\beta H_\mathrm{A} S + \sum_\mathbf{k} (g\beta H_\mathrm{A} - 2JSn)(c_\mathbf{k}^+ c_\mathbf{k} + d_\mathbf{k}^+ d_\mathbf{k})$$
$$- 2JS\gamma_\mathbf{k}(c_\mathbf{k}^+ d_\mathbf{k}^+ + c_\mathbf{k} d_\mathbf{k}), \qquad (6.203)$$

where

$$\gamma_\mathbf{k} = \sum_\mathbf{r} \exp(i\mathbf{k}\cdot\mathbf{r}),$$

$\mathbf{r}$ being a vector from an atom to a nearest neighbor. Terms of higher order in operators $c_\mathbf{k}$ and $d_\mathbf{k}$ have been omitted in the above Hamiltonian.

In order to find the eigenstates of the Hamiltonian, the following operators are introduced:

$$A_\mathbf{k} = u_\mathbf{k} c_\mathbf{k} - v_\mathbf{k} d_\mathbf{k}^+, \qquad A_\mathbf{k}^+ = u_\mathbf{k} c_\mathbf{k}^+ - v_\mathbf{k} d_\mathbf{k},$$
$$B_\mathbf{k} = u_\mathbf{k} d_\mathbf{k} - v_\mathbf{k} c_\mathbf{k}^+, \qquad B_\mathbf{k}^+ = u_\mathbf{k} d_\mathbf{k}^+ - v_\mathbf{k} c_\mathbf{k}, \qquad (6.204)$$

where $u_{\mathbf{k}}$ and $v_{\mathbf{k}}$ are real numbers satisfying the condition $u_{\mathbf{k}}^2 - v_{\mathbf{k}}^2 = 1$. The Hamiltonian becomes

$$\begin{aligned} H = 2NnJS^2 &= 2Ng\beta H_A S - N(g\beta H_A - 2JnS) \\ &\quad + \sum_{\mathbf{k}} \hbar\omega_{\mathbf{k}}(A_{\mathbf{k}}^+ A_{\mathbf{k}} + B_{\mathbf{k}}^+ B_{\mathbf{k}} + 1), \end{aligned} \tag{6.205}$$

in which

$$(\hbar\omega_{\mathbf{k}})^2 = (g\beta H_A - 2JnS)^2 - (2JS)^2\gamma_{\mathbf{k}}^2.$$

It is to be remembered that J is negative. We see that the eigenstates of H are eigenstates of $(A_{\mathbf{k}}^+ A_{\mathbf{k}} + B_{\mathbf{k}}^+ B_{\mathbf{k}})$. The eigenvalues of H are given by

$$2NnJS(S+1) - 2Ng\beta H_A(S + \tfrac{1}{2}) + \sum_{\mathbf{k}} \hbar\omega_{\mathbf{k}}(n_{\mathbf{k}} + \tfrac{1}{2}), \tag{6.206}$$

in which $n_{\mathbf{k}}$ is a positive integer and the energy quantum $\hbar\omega_{\mathbf{k}}$ is called the antiferromagnetic magnon.

The operator of the z component of all spins on sublattice a is

$$\begin{aligned} S_z^a &= \sum_i S_{iz}^a = NS - \sum_{\mathbf{k}} c_{\mathbf{k}}^+ c_{\mathbf{k}} \\ &= NS - \sum_{\mathbf{k}} [u_{\mathbf{k}}^2 A_{\mathbf{k}}^+ A_{\mathbf{k}} + v_{\mathbf{k}}^2 B_{\mathbf{k}} B_{\mathbf{k}}^+ + u_{\mathbf{k}} v_{\mathbf{k}} (A_{\mathbf{k}} B_{\mathbf{k}} + A_{\mathbf{k}}^+ B_{\mathbf{k}}^+)]. \end{aligned} \tag{6.207}$$

The operator is not diagonal for the eigenstates of H. Consider the ground state, which has all $n_{\mathbf{k}}$'s equal to zero. In the approximation of neglecting H_A, the energy of the state is

$$E_0 = 2NnJS\left\{S + N^{-1}\sum_{\mathbf{k}}\left[1 - \left(1 - \frac{\gamma_{\mathbf{k}}^2}{n^2}\right)^{1/2}\right]\right\}, \tag{6.208}$$

and the expectation value of S_z^a is

$$\langle S_z^a \rangle = NS - \frac{1}{2}\sum_{\mathbf{k}}\left[\left(1 - \frac{\gamma_{\mathbf{k}}^2}{n^2}\right)^{-1/2} - 1\right]. \tag{6.209}$$

We see that $\langle S_z^a \rangle$ is less than NS; even in the ground state, the spins of a sublattice are not perfectly aligned.

c. Interaction involving Conduction Electrons

Indirect Exchange of Localized Spins. The preceding section dealt exclusively with localized spins. The formation of a localized magnetic moment in the presence of conduction electrons has been the subject of many theoretical treatments, beginning with that of P. W. Anderson. Local moments, should they exist, interact via the itinerant electrons; such interaction was first treated by M. A. Ruderman and C. Kittel by considering nuclear moments. The indirect interaction may be important if the concentration of itinerant electrons is large, such as in a metal, while the separation between localized spins is too large for the direct exchange to be significant. Such a situation occurs in rare earth metals and in nonmagnetic metals containing a magnetic impurity. The perturbation due to the interaction of an ionic spin S_{i} with the itinerant electrons can be expressed in the following form:

$$H_{\mathrm{i}} = -\sum_{\mathbf{k}} A(\mathbf{r}_{\mathbf{k}\sigma} - \mathbf{R}_{\mathrm{i}})\mathbf{S}_{\mathbf{k}} \cdot \mathbf{S}_{\mathrm{i}}, \qquad (6.210)$$

where $\mathbf{k}$ is the wavevector and σ indicates the spinor of an itinerant electron. The energy perturbation of second order, involving two ionic spins, is given by

$$E''_{\mathrm{ii}'} = 2\sum_{\mathbf{k}\sigma}\sum_{\mathbf{k}'\sigma'} \frac{\langle \mathbf{k}\sigma | H_{\mathrm{i}'} | \mathbf{k}'\sigma' \rangle \langle \mathbf{k}'\sigma' | H_{\mathrm{i}} | \mathbf{k}\sigma \rangle}{\varepsilon_{\mathbf{k}} - \varepsilon_{\mathbf{k}'}}. \qquad (6.211)$$

The factor 2 takes into account the identical contribution given by switching H_{i} and $H_{\mathrm{i}'}$ in the above expression. Let $|0\rangle$ denote the unperturbed state of the system. Straightforward calculation gives

$$\begin{aligned} E''_{\mathrm{ii}'} &= 2\sum_{\sigma} \langle 0 | (\mathbf{S}_{\mathrm{i}'} \cdot \mathbf{S}_{\sigma})(\mathbf{S}_{\mathrm{i}} \cdot \mathbf{S}_{\sigma}) | 0 \rangle \\ &\quad \times \left\{ \sum_{\mathbf{k}'} \exp[-i(\mathbf{k} - \mathbf{k}') \cdot (\mathbf{R}_{\mathrm{i}'} - \mathbf{R}_{\mathrm{i}})] \sum_{\mathbf{k}} \frac{J(\mathbf{k},\mathbf{k}')J(\mathbf{k}',\mathbf{k})}{\varepsilon_{\mathbf{k}} - \varepsilon_{\mathbf{k}'}} \right\} \\ &= \langle 0 | \mathbf{S}_{\mathrm{i}'} \cdot \mathbf{S}_{\mathrm{i}} | 0 \rangle \{ \ \}. \end{aligned} \qquad (6.212)$$

In the above expression,

$$J(\mathbf{k}',\mathbf{k}) = \int d\mathbf{r}\, \varphi^*_{\mathbf{k}'}(\mathbf{r}) A(\mathbf{r}) \varphi_{\mathbf{k}}(\mathbf{r}), \qquad (6.213)$$

$\varphi_{\mathbf{k}}(\mathbf{r})$ being a Bloch function for itinerant electrons. The summations in the curly brackets are over occupied states $\mathbf{k}$ and unoccupied states $\mathbf{k}'$, respectively. The dependence of $\{ \ \}$ on $\mathbf{R}_{\mathrm{i}'} - \mathbf{R}_{\mathrm{i}}$ shows that $E''_{\mathrm{ii}'}$ of the indirect exchange oscillates with variation of ion separation. The exchange can lead to a large variety of ordered spin structures, including spirals.

The group of itinerant electrons is the intermediary of the indirect interaction. It is apparent, indeed it can be shown, that the oscillatory nature of indirect interaction comes from the consequence that a localized spin produces spin polarization of the group of electrons, and the polarization oscillates with distance from the localized spin.

The Kondo Effect. The electrical resistivity of alloy metals with a dilute magnetic component shows a minimum at low temperature. J. Kondo showed that the phenomenon results from the scattering of conduction electrons by local moments. The interaction H_{i} (6.210) produces the transition of an electron from an occupied state $|\mathbf{k}\sigma\rangle$ to a vacant state $|\mathbf{k}'\sigma'\rangle$. The transition matrix element to the second order of perturbation is

$$\langle\mathbf{k}'\sigma'|M|\mathbf{k}\sigma\rangle = \langle\mathbf{k}'\sigma'|\left(H_{\mathrm{i}} + \sum_{\mathbf{k}''\sigma''} \frac{H_{\mathrm{i}}|\mathbf{k}''\sigma''\rangle\langle\mathbf{k}''\sigma''|H_{\mathrm{i}}}{\varepsilon_{\mathbf{k}} - \varepsilon_{\mathbf{k}''}}\right)|\mathbf{k}\sigma\rangle. \quad (6.214)$$

Take the factor $J(\mathbf{k}',\mathbf{k})$ of H_{i} to be J independent of $\mathbf{k}$ and $\mathbf{k}'$. Furthermore, the significance of such transitions can be clearly and more simply demonstrated by considering only transitions that do not change the electron spin between the initial and the final states, that is, $|\mathbf{k}\sigma\rangle \to |\mathbf{k}'\sigma\rangle$, but do include the option $\sigma'' \neq \sigma = \sigma'$. The rate of transition is proportional to the transition matrix element squared, which to the third order in J is

$$|\langle\mathbf{k}'+|M|\mathbf{k}+\rangle|^2 = \left(\frac{J}{2}\right)^2 (S_{iz})^2 \left(1 - \frac{1}{2}\sum_{\mathbf{k}''} \frac{2f(\mathbf{k}'') - 1}{\varepsilon_{\mathbf{k}} - \varepsilon_{\mathbf{k}''}}\right). \quad (6.215)$$

The expression contains the distribution function $f(\mathbf{k}'')$ because the electron occupation of the intermediate state is involved in the second-order transition.

It can be seen that the important second-order transitions are associated with $\varepsilon(\mathbf{k}) \gtrless \varepsilon_{\mathrm{F}}$ and $\varepsilon(\mathbf{k}'') \lessgtr \varepsilon_{\mathrm{F}}$. The transitions become divergent at $T = 0$. Calculations show that the transitions contribute a term in the resistivity of the following form:

$$\rho \propto J \ln(kT/\varepsilon_{\mathrm{F}}). \quad (6.216)$$

With a negative J, the contribution leads to an increase in resistivity with decreasing temperature, producing a minimum of resistivity at low temperature, the so-called Kondo effect.

d. Exchange Interaction of Itinerant Electrons

The exchange interaction of itinerant electrons may lead to spontaneous magnetization of the electron spins. In the Fock Hamiltonian H^{F}, Eq. (3.5), the exchange interaction is represented by the operator A. Introduce the second-quantization creation and annihilation operators, c_α^+ and c_α, the subscript of which indicates the one-electron state including spin. In terms of these operators, the Slater determinant Ψ takes on the concise form

$$|\Psi\rangle = c_\nu^+ \cdots c_\beta^+ c_\alpha^+ |0\rangle; \tag{6.217}$$

$|0\rangle$ represents the vacuum, $\langle 0 \mid 0\rangle = 1$. The number of various c^+'s is equal to the number of electrons of the system. The complex-conjugate wavefunction Ψ^* has the form

$$\langle \Psi^* | = \langle 0 \mid c_\alpha c_\beta \cdots c_\nu. \tag{6.218}$$

Evidently,

$$c_\mu |0\rangle = 0 \qquad \text{and} \qquad \langle 0| c_\mu^+ = 0 \tag{6.219}$$

for any μ. The following commutation relations of the operators represent the antisymmetry of the determinant:

$$c_\alpha^+ c_\beta^+ + c_\beta^+ c_\alpha^+ = 0, \quad c_\alpha c_\beta + c_\beta c_\alpha = 0, \quad c_\alpha^+ c_\beta + c_\beta c_\alpha^+ = \delta_{\alpha\beta}. \tag{6.220}$$

It is easy to show by applying the commutation relations that

$$\langle \Psi^* \mid \Psi\rangle = \langle 0| c_\alpha c_\beta \cdots c_\nu c_\nu^+ \cdots c_\beta^+ c_\alpha^+ |0\rangle = \langle 0 \mid 0\rangle = 1, \tag{6.221}$$

that is, the determinant Ψ is normalized.

Consider the simple case for which only one energy band need be considered. A one-electron state may be written as $|\kappa\rangle$, where κ stands for a combination of the Bloch wavevector and electron spin. In terms of the operators c^+ and c, the electron–electron interaction terms of the Fock Hamiltonian, the last two terms of (3.5), are, respectively,

$$\tfrac{1}{2} \sum_{\kappa} \sum_{\kappa'}{}' \langle \kappa', \kappa | V_{12} | \kappa, \kappa' \rangle c_{\kappa'}^+ c_\kappa^+ c_\kappa c_{\kappa'}$$

and

$$\tfrac{1}{2} \sum_{\kappa} \sum_{\kappa'}{}' \langle \kappa', \kappa | V_{12} | \kappa', \kappa \rangle c_{\kappa'}^+ c_\kappa^+ c_{\kappa'} c_\kappa, \tag{6.222}$$

where

$$\langle \kappa', \kappa | V_{12} | \kappa', \kappa \rangle = \int \varphi_{\kappa'}^*(1) \varphi_\kappa^*(2) \frac{e^2}{\mathbf{r}_1 - \mathbf{r}_2} \varphi_{\kappa'}(2) \varphi_\kappa(1)\, d\tau_1\, d\tau_2. \tag{6.223}$$

The first term is the interaction that is also present in the Hartree approximation; it is independent of the spins of interacting electrons. The second term represents the operator A of the exchange interaction. It depends on the spins of the φ's, giving the correlation of electrons of the same spin. We will consider the effect of A on the spin configuration of the many-electron system, keeping in mind that correlation of electrons with opposite spins should also be taken into account.

The contribution of exchange interaction to the energy of the system is

$$\tfrac{1}{2}\langle\Psi|\sum_{\kappa}\sum_{\kappa'}\langle\kappa',\kappa|V_{12}|\kappa',\kappa\rangle c_{\kappa'}^{+}c_{\kappa}^{+}c_{\kappa'}c_{\kappa}|\Psi\rangle. \tag{6.224}$$

$c_{\kappa'}^{+}c_{\kappa}^{+}c_{\kappa'}c_{\kappa}$ gives $-n_{\kappa'}n_{\kappa}$, where n is occupation of the one-electron state. If the matrix element of V_{12} is positive, the exchange interaction lowers the energy of the system; that is, the interaction among electrons of the same spin favors the condition of one common spin, leading to ferromagnetism of itinerant electrons. Actually, the collective spin structure is a complicated problem. The Fock Hamiltonian is an approximation based on one Slater determinant of one-electron functions, and it does not take into account the correlation of electrons with opposite spins. Many treatments have been made using various assumptions. It has been shown that ferromagnetism and spin waves may be given by itinerant electrons. Furthermore, under certain conditions, the ground state may consist of a spin density wave that is oscillatory spin density of uniformly distributed electrons; such a case constitutes antiferromagnetism.

5. MOLECULAR FIELD THEORY

The exchange interaction of electrons, which is crucial for magnetic properties, is a very difficult problem for theoretical treatment. The simplest case is that considered in Section 4.a, for which the operators S in (6.188) are those of individual atoms. In terms of electronic energy bands, the case applies to insulators in which the electron-occupied energy bands are quite narrow, for example, EuS and MnF_2. In principle, all electrons should be considered "itinerant" except for those in very narrow energy bands. It has been pointed out that (6.224) is only an approximate expression of the exchange interaction of itinerant electrons. Even in this expression, approximations regarding $\langle\kappa'\kappa|V_{12}|\kappa'\kappa\rangle$ are necessary for practicable calculation. We shall not consider the numerous treatments involving various refinements.

The simple theory given by Pierre Weiss in 1903 has often been applied in approximate considerations. It deals with ferromagnetism on the basis of atomic (nearly localized electron) approximation. The effect of surrounding atoms is represented by a magnetic field $\mathbf{H}_\mathrm{L}$ that has been referred to as a local field or a molecular field:

$$\mathbf{H}_\mathrm{L} = \alpha \mathbf{M}, \tag{6.225}$$

where $\mathbf{M}$ is macroscopic magnetization. The Weiss theory preceded the concept of exchange interaction, and α was taken to be an empirical parameter. We see now that the exchange interaction involving an atom i has the form

$$-g\beta \mathbf{S}_i \cdot \frac{1}{g\beta} \sum_j 2J_{ij}\mathbf{S}_j = -g\beta \mathbf{S}_i \cdot \left(\frac{2Jn}{g\beta} \mathbf{S} \right)_i, \tag{6.226}$$

where n is the number of nearest neighbors of the atom and $J\mathbf{S}$ is the average over the neighbors. In the case where n and $J\mathbf{S}$ are the same for every atom, the effect of exchange interaction is equivalent to the effect of a magnetic field $\mathbf{H}_L$ defined by

$$\mathbf{H}_L = \frac{2Jn}{g\beta} \mathbf{S} = \frac{2Jn}{Ng^2\beta^2} Ng\beta \mathbf{S}, \tag{6.227}$$

where N is the number of atoms per unit volume. When the orbital angular momentum of each atom has $L = 0$, $Ng\beta\mathbf{S}$ is the magnetic moment per unit volume, $\mathbf{M}$. The last equation relates the phenomenological parameter α with the exchange interaction.

With reference to Section VI.D.3, the magnetic moment $\mathbf{M}_\mathrm{a}$ of an atom with scalar susceptibility χ_a is parallel to the magnetic field experienced by the atom. Under an applied field $\mathbf{H}$, the field experienced by an atom is $\mathbf{H} + \alpha\mathbf{M}$. The magnetic moment per unit volume of the solid, $\mathbf{M} = n\mathbf{M}_a$, is then parallel to $\mathbf{H} + \alpha\mathbf{M}$, and hence $\mathbf{M} \parallel \mathbf{H}$. We then have an equation for the magnitude of $\mathbf{M}$,

$$M = N\chi_\mathrm{a}(H + \alpha M) = N\beta g J B_J \left(\frac{\beta g (H + \alpha M)}{kT} \right), \tag{6.228}$$

upon using expression (6.178) for χ_a. Here J is the quantum number of atomic total angular momentum, not to be confused with J of the exchange interaction, and $B_J(\eta)$ is the Brillouin function.

The Brillouin function $B_J(\eta)$ increases smoothly with increasing argument η, from 0 at $\eta = 0$ to a saturation value of unity. With the notations

$$M_0 = N\beta gJ \qquad \text{and} \qquad \eta = \frac{\beta g}{kT}H + \left(\frac{\alpha\beta g M_0}{kT}\right)\frac{M}{M_0}, \tag{6.229}$$

(6.228) can be written as

$$\frac{M}{M_0} = B_J(\eta). \tag{6.230}$$

This implicit equation for M/M_0 can be solved graphically. Let the y axis of a rectangular coordinate system be the axis of M/M_0 and the x axis be the axis of η, that is, $\eta = x + \beta gH/kT$. The curve $B_J(\eta)$ extends from $B_J(-\beta gH/kT) = 0$ to $B_J(\infty) = 1$. Draw a straight line of slope $(\alpha\beta g M_0/kT)^{-1}$ from the point $(x = 0,\ y = 0)$. The ordinate of the intersection of this straight line with the $B_J(\eta)$ curve gives M/M_0. Straight lines for various T lead to the temperature dependence of M/M_0 for the applied H. As $T \to 0$, $M/M_0 \to 1$ irrespective of H; M_0 is the maximum magnetization.

An M existing in the absence of an applied field H is spontaneous magnetization. With $H = 0$, $B_J(\eta)$ begins at $\eta = x = 0$. The slope $\partial B_J(\eta)/\partial\eta$ of the Brillouin function decreases steadily with increasing η, from a maximum value of $(J+1)/3$ near $\eta = 0$. In order for one of the above-mentioned straight lines to intersect the curve $B_J(x)$, its slope must be smaller than $(J+1)/3$:

$$\left(\frac{\alpha\beta g M_0}{kT}\right)^{-1} \leq \tfrac{1}{3}(J+1).$$

The temperature

$$T_c = \frac{\alpha\beta g M_0}{k}\frac{J+1}{3} = N\beta g\frac{\alpha\beta g}{k}\frac{J(J+1)}{3} \tag{6.231}$$

at which the spontaneous magnetization M vanishes is the Curie temperature.

Consider the differential magnetic susceptibility

$$\chi \equiv \left(\frac{dM}{dH}\right)_{H=0}. \tag{6.232}$$

Substituting in expression (6.228) for M, we get

$$\begin{aligned}\chi &= N\beta gJ\frac{\beta g}{kT}\left[\left(1+\alpha\frac{dM}{dH}\right)\frac{dB_J(\eta)}{d\eta}\right]_{H=0}\\ &= N\beta gJ\frac{\beta g}{kT}(1+\alpha\chi)\left(\frac{dB_J}{d\eta}\right)_{H=0}. \end{aligned}\tag{6.233}$$

At a temperature $T > T_c$, there is no spontaneous magnetization and therefore $\eta = 0$ for $H = 0$. We have, to the first order of H,

$$\left(\frac{dB_J(\eta)}{d\eta}\right)_{H=0} \approx \left(\frac{dB_J(\eta)}{d\eta}\right)_{\eta} = \frac{J+1}{3}.$$

Combining the last two equations, we get the Curie–Weiss law:

$$\chi = \frac{T_c}{\alpha(T-T_c)}. \tag{6.234}$$

Antiferromagnetism. Consider the simple case where the magnetic structure of the crystal can be divided into two sublattices. The nearest neighbors of an ion of one sublattice are ions of the other sublattice. For an antiferromagnet, the parameter α in (6.225) is negative, whereas it is positive in the case of a ferromagnet. There are two simultaneous equations in place of (6.230):

$$\frac{M_1}{M_0} = B_J\left(\frac{\beta g}{kT}H + \frac{\alpha\beta g M_0}{kT}\frac{M_2}{M_0}\right)$$

and

$$\frac{M_2}{M_0} = B_J\left(\frac{\beta g}{kT}H + \frac{\alpha\beta g M_0}{kT}\frac{M_1}{M_0}\right). \tag{6.235}$$

The subscripts 1 and 2 are sublattice indices. In the notation $M_0 = N\beta gJ$, N is the ion concentration of each sublattice. Consider the situation of no externally applied field, $H = 0$. The parameter α being negative, the spontaneous magnetizations M_1 and M_2 are opposite in sign. The symmetry of the two equations shows that $M_1 = -M_2$. Thus we have

$$\frac{M_1}{M_0} = B_J\left(-\frac{\alpha\beta g M_0}{kT}\frac{M_1}{M_0}\right) = B_J\left(\frac{|\alpha|\beta g M_0}{kT}\frac{M_1}{M_0}\right) \tag{6.236}$$

and a similar equation for M_2/M_0. Each equation is similar to the equation for a ferromagnet. We see from similar considerations that with increasing temperature M_1 and M_2 vanish at the temperature

$$T_N = N\frac{|\alpha|\beta g M_0}{k}\frac{J+1}{3} = N\beta g\frac{|\alpha|\beta g}{k}\frac{J(J+1)}{3}, \tag{6.237}$$

which is called the Néel temperature.

Consider the magnetic susceptibility of the antiferromagnet:

$$\chi = \left(\frac{dM}{dH}\right)_{H=0} = \left(\frac{d(M_1+M_2)}{dH}\right)_{H=0}. \tag{6.238}$$

For temperatures not too far from T_N, we may write

$$\frac{M_1}{M_0} \sim \frac{J+1}{3}\left(\frac{\beta g}{kT}H + \alpha M_0 \frac{\beta g}{kT}\frac{M_2}{M_0}\right)$$

and

$$\frac{M_2}{M_0} \sim \frac{J+1}{3}\left(\frac{\beta g}{kT}H + \alpha M_0 \frac{\beta g}{kT}\frac{M_1}{M_0}\right). \tag{6.239}$$

Differentiating the sum of the two equations, we get

$$\left(\frac{dM}{dH}\right)_{H=0} = M_0 \frac{J+1}{3}\frac{\beta g}{kT}\left[2 + \alpha\left(\frac{dM}{dH}\right)_{H=0}\right],$$

which gives

$$\chi = \frac{2T_N/|\alpha|}{T+T_N}. \tag{6.240}$$

The Ising Model. Molecular field theory is highly simplified for the treatment of a system of moments coupled by exchange interaction. The use of macroscopic magnetization to represent the interactions among the localized moments does not properly account for the possible states of the system, for example, states containing various numbers of spin waves. The inadequacy of the theory is particularly significant for the approach of spontaneous magnetization M toward M_0 at $T=0$ and for the range of $M \sim 0$. Extensive studies have been made on the Ising model, which was introduced in 1925. The model for a system of spins considers only the spin components along a particular axis. This simplification makes exact solutions obtainable for two-dimensional systems of simple structures. The results obtained are useful as reference in many respects.

Let the component of **S** along the particular axis be denoted simply by S. The Ising model restricts S to having two values, ± 1. Only interactions of neighbors are considered. For a two-dimensional, rectangular lattice, the problem involves the Hamiltonian

$$\begin{aligned} H_{ex} - \beta g H \sum_i S_i = & -2J_1 \sum_n S_{n,m}S_{n+1,m} \\ & - 2J_2 \sum_m S_{n,m}S_{n,m+1} - \beta g H \sum_i S_i, \end{aligned} \tag{6.241}$$

where H is the applied static magnetic field; n and m refer to the nth row and the mth column, respectively, of the lattice; J_1 is the exchange parameter of two neighbors in a row; and J_2 is the parameter of two neighbors in a column of the lattice. For a ferromagnetic system, the exact expression derived for the spontaneous magnetization $M(H=0)$ is

$$\frac{M}{M_0} = \begin{cases} [1-\operatorname{csch}^2(J_1/kT)\operatorname{csch}^2(J_2/kT)]^{1/8} & \text{for} \quad T < T_c \\ 0 & \text{for} \quad T > T_c \end{cases}, \tag{6.242}$$

where T_c is given by

$$\sinh(J_1/kT_c)\sinh(J_2/kT_c) = 1 \tag{6.243}$$

and M_0 is the maximum possible value of magnetization. Consider the case of $J_1 = J_2$. The temperature T_c is given by $kT_c/J = 1.135$, whereas the Curie temperature in molecular field theory would be given by $kT_c/J = 2$, according to (6.231) with $J = \frac{1}{2}$ and α given by (6.227). The magnetization approaches zero with infinite slope as T approaches T_c from below, as $M/M_0 \propto (T_c - T)^{1/8}$. According to molecular field theory, on the other hand, the spontaneous magnetization approaches zero with infinite slope as T increases toward the Curie temperature T_c, but according to $M/M_0 \propto (T_c - T)^{1/2}$. The above comparison shows the differences in T_c and in the temperature variation in M predicted by the two theories.

E. SUPERCONDUCTIVITY

1. ELECTRON–ELECTRON INTERACTION INVOLVING PHONONS

Superconductivity was discovered in 1911. Its theoretical treatment, the Bardeen–Cooper–Schrieffer (BCS) theory, was developed in the 1950s. The idea that electron–phonon interaction is involved occurred to H. Frölich in 1950 and it was confirmed by the observation of the isotope effect, a dependence of the superconducting temperature upon the isotope mass of nuclei in the metal.

Consider the following Hamiltonian expressed in terms of the creation and annihilation operators c^+ and c of electrons and a^+ and a of phonons:

$$H = H_0 + H' = \sum_{\mathbf{q}} \hbar\omega_{\mathbf{q}} a_{\mathbf{q}}^+ a_{\mathbf{q}} + \sum_{\mathbf{k}s} \epsilon_{\mathbf{k}} c_{\mathbf{k}s}^+ c_{\mathbf{k}s} + H'. \tag{6.244}$$

where $\mathbf{q}$ and $\hbar\omega_{\mathbf{q}}$ are the wavevector and energy, respectively, of a phonon; $\mathbf{k}$ and $\mathbf{s}$ are the wavevector and spin, respectively, of a one-electron Bloch state; and H' represents the electron–phonon interaction. Taking the deformation potential interaction outlined in Section V.1.a, we have

$$H' = i\sum_{\mathbf{skq}} \left(\frac{\hbar}{N2M\omega_{\mathbf{q}}}\right)^{1/2} Cq(c^{+}_{k+q,\mathbf{s}}c_{\mathbf{k},\mathbf{s}}a_{\mathbf{q}} - c^{+}_{\mathbf{k}-\mathbf{q},\mathbf{s}}c_{\mathbf{k},\mathbf{s}}a^{+}_{\mathbf{q}})$$
$$\equiv i\sum_{\mathbf{skq}} D_{\mathbf{q}}c^{+}_{\mathbf{k}+\mathbf{q},\mathbf{s}}c_{\mathbf{k},\mathbf{s}}(a_{\mathbf{q}} - a^{+}_{-\mathbf{q}}). \tag{6.245}$$

Make a unitary transformation of the Hamiltonian:

$$H_{\mathrm{T}} \equiv e^{-S}He^{S} = H + [H,S] + \tfrac{1}{2}[[H,S],S] + \cdots. \tag{6.246}$$

Using an operator S that satisfies

$$[H_0, S] = -H', \tag{6.247}$$

we get a transformed Hamiltonian

$$H_{\mathrm{T}} = H_0 + \tfrac{1}{2}[H', S] + \cdots, \tag{6.248}$$

which does not have terms linear in H'. The matrix element of S connecting two eigenstates of H_0, $|m\rangle$ and $|n\rangle$, is given by

$$\langle n|S|m\rangle = \frac{\langle n|H'|m\rangle}{E_m - E_n}. \tag{6.249}$$

A matrix element of H' exists only between two states that differ in the quantum number $n_{\mathbf{q}}$ of a phonon mode $\mathbf{q}$ by ± 1, because

$$\langle n_{\mathbf{q}} - 1|a_{\mathbf{q}}|n_{\mathbf{q}}\rangle = n_{\mathbf{q}}^{1/2} \qquad \text{and} \qquad \langle n_{\mathbf{q}} + 1|a^{+}_{\mathbf{q}}|n_{\mathbf{q}}\rangle = (n_{\mathbf{q}} + 1)^{1/2}.$$

Hence the operator S has the matrix elements

$$\langle \mathbf{k}+\mathbf{q}, \mathbf{s}; n_{\mathbf{q}} - 1|S|\mathbf{k}, \mathbf{s}; n_{\mathbf{q}}\rangle = \frac{iD_{\mathbf{q}}c^{+}_{\mathbf{k}+\mathbf{q},\mathbf{s}}c_{\mathbf{k},\mathbf{s}}n_{\mathbf{q}}^{1/2}}{\epsilon_{\mathbf{k}} - \epsilon_{\mathbf{k}+\mathbf{q}} + \hbar\omega_{\mathbf{q}}}$$

and

$$\langle \mathbf{k}-\mathbf{q}, \mathbf{s}; n_{\mathbf{q}} + 1|S|\mathbf{k}, \mathbf{s}; n_{\mathbf{q}}\rangle = \frac{-iD_{\mathbf{q}}c^{+}_{\mathbf{k}-\mathbf{q},\mathbf{s}}c_{\mathbf{k},\mathbf{s}}(n_{\mathbf{q}} + 1)^{1/2}}{\epsilon_{\mathbf{k}} - \epsilon_{\mathbf{k}-\mathbf{q}} - \hbar\omega_{\mathbf{q}}}. \tag{6.250}$$

In terms of the matrix elements of H' and S, we have for a fixed set of $n_{\mathbf{q}}$

$$\langle|\tfrac{1}{2}[H'S]|\rangle = \tfrac{1}{2}\sum_{\mathbf{ksk's'q}}\{\langle\mathbf{k'}\pm\mathbf{q},\mathbf{s'};n_{\mathbf{q}}|H'|\mathbf{k's'};n_{\mathbf{q}}\pm 1\rangle \times\langle\mathbf{k}\mp\mathbf{q},\mathbf{s};n_{\mathbf{q}}\pm 1|S|\mathbf{ks};n_{\mathbf{q}}\rangle - \langle\mathbf{k'}\pm\mathbf{q},\mathbf{s'};n_{\mathbf{q}}|S|\mathbf{k's'};n_{\mathbf{q}}\pm 1\rangle \times\langle\mathbf{k}\mp\mathbf{q},\mathbf{s};n_{\mathbf{q}}\pm 1|H'|\mathbf{ks};n_{\mathbf{q}}\rangle\}.$$

This will be denoted by H_{I}. It represents electron–electron interaction by means of phonons. Substituting the expressions of the matrix elements, we get

$$H_{\mathrm{I}} = \sum_{\mathbf{ksk's'q}} |D_{\mathbf{q}}|^2\hbar\omega_{\mathbf{q}}[(\epsilon_{\mathbf{k}}-\epsilon_{\mathbf{k-q}})^2-(\hbar\omega_{\mathbf{q}})^2]^{-1}c^{+}_{\mathbf{k-q,s}}c^{+}_{\mathbf{k'+q,s'}}c_{\mathbf{k's'}}c_{\mathbf{ks}}. \tag{6.251}$$

The combination $c_{\mathbf{k's'}}c_{\mathbf{ks}}$ of operators destroys a pair of electrons, and the combination $c^{+}_{\mathbf{k-q,s}}c^{+}_{\mathbf{k'+q,s'}}$ adds a pair of electrons, in different one-electron states.

In the absence of H_{I}, the one-electron states are nearly filled up to the Fermi energy ϵ_{F}, and the states with $\epsilon > \epsilon_{\mathrm{F}}$ are practically unoccupied by electrons. The replacement of $|\mathbf{ks}\rangle$ and $|\mathbf{k's'}\rangle$ by $|\mathbf{k}\pm\mathbf{q},\mathbf{s}\rangle|\mathbf{k'}\pm\mathbf{q},\mathbf{s'}\rangle$ may be expected to be important mainly for

$$\epsilon_{\mathbf{k}},\epsilon_{\mathbf{k'}} > (\epsilon_{\mathrm{F}}-\hbar\omega_{\mathrm{m}}) \qquad \text{and} \qquad \epsilon_{\mathbf{k}\pm\mathbf{q}},\epsilon_{\mathbf{k'}\pm\mathbf{q}} < (\epsilon_{\mathrm{F}}+\hbar\omega_{\mathrm{m}}), \tag{6.252}$$

where $\hbar\omega_{\mathrm{m}}$ is the maximum phonon energy; usually $\hbar\omega_{\mathrm{m}} \ll \epsilon_{\mathrm{F}}$.

Write the expression of H_{I} as

$$H_{\mathrm{I}} = \sum_{\mathbf{ksk's'q}} V_{\mathbf{kk'q}}c^{+}_{\mathbf{k-q,s}}c^{+}_{\mathbf{k'+q,s'}}c_{\mathbf{k's'}}c_{\mathbf{ks}}. \tag{6.253}$$

The coefficients V with $|\epsilon_{\mathbf{k}\pm\mathbf{q}}-\epsilon_{\mathbf{k}}| > \hbar\omega_{\mathrm{m}}$ are positive, and the others are negative. The negative part of H_{I}, the attractive interaction, is the cause of the superconducting state, and it will be the focus of our attention.

Cooper Pairs. It was shown by L. Cooper that the formation of electron pairs, Cooper pairs, bound by the exchange of phonons has an unusual effect on the electron gas. The result pointed out the direction for the development of the BCS theory of superconductivity.

Consider two electrons in a degenerate free-electron system. The two one-electron states, 1 and 2, of the pair are to have antiparallel spins, and therefore the spatial wavefunction of the pair is symmetric under permutation of electrons:

$$\psi = \exp[i(\mathbf{k}_1 \cdot \mathbf{r}_1 + \mathbf{k}_2 \cdot \mathbf{r}_2)] \tag{6.254}$$

is normalized per unit volume. The "center-of-mass" and "relative-motion" coordinates of the pair are

$$\mathbf{R} = (\mathbf{r}_1 + \mathbf{r}_2)/2, \qquad \mathbf{K} = \mathbf{k}_1 + \mathbf{k}_2,$$

and

$$\mathbf{r} = \mathbf{r}_1 - \mathbf{r}_2, \qquad \boldsymbol{\kappa} = (\mathbf{k}_1 - \mathbf{k}_2)/2. \tag{6.255}$$

In terms of these quantities,

$$\psi = \exp[i(\mathbf{K} \cdot \mathbf{R} + \boldsymbol{\kappa} \cdot \mathbf{r})], \tag{6.256}$$

with an energy of the state equal to $(\hbar^2 m)(K^2/4 + \kappa^2)$. To consider pairs of low energy, we take $K = 0$ associated with $\boldsymbol{\kappa} = \mathbf{k}_1 = -\mathbf{k}_2$.

In the presence of electron–electron interaction H_I, the wavefunction of a pair may be expressed as

$$\psi = \sum_{\kappa} a_\kappa \exp(i\boldsymbol{\kappa} \cdot \mathbf{r}). \tag{6.257}$$

For a pair with energy eigenvalue E, we get from $(H - E)\psi = 0$ the following secular equation for the coefficients a_κ:

$$\int d\mathbf{r} \exp(-i\boldsymbol{\kappa} \cdot \mathbf{r})(H - E) \sum_{\kappa'} a_{\kappa'} \exp(i\boldsymbol{\kappa}' \cdot \mathbf{r}) = 0. \tag{6.258}$$

The energy of a pair having $\boldsymbol{\kappa}$ in the absence of interaction is $E^0_\kappa = \hbar^2 k^2/m$. It follows from Eq. (6.258) that

$$(E^0_\kappa - E)a_\kappa + \sum_{\kappa'} a_{\kappa'} \langle \mathbf{k}, -\mathbf{k}|H_\mathrm{I}|\mathbf{k}', -\mathbf{k}'\rangle = 0, \tag{6.259}$$

where

$$\mathbf{k}' = \mathbf{k} - \mathbf{q} \qquad \text{and} \qquad -\mathbf{k}' = -(\mathbf{k} - \mathbf{q}).$$

Let $\rho(E^0)$ denote the density of pair states $(\mathbf{k}, -\mathbf{k})$. The preceding equation can be written as

$$(E^0 - E)a(E^0) + \int dE^{0'} \rho(E^{0'}) a(E^{0'}) \langle E^0|H_\mathrm{I}|E^{0'}\rangle = 0. \tag{6.260}$$

Replace the matrix element of H_I by its average value V put in front of the integral. Consider a pair state $|\psi\rangle$ made of one-electron states with energies between ϵ_F and $\epsilon_\mathrm{m} = \epsilon_\mathrm{F} + \hbar\omega_\mathrm{m}$,

$$(E^0 - E)a(E^0) = V \int_{2\epsilon_\mathrm{F}}^{2\epsilon_\mathrm{m}} dE^{0'} \rho(E^{0'}) a(E^{0'}) \equiv C. \tag{6.261}$$

V is assumed to be positive and independent of E^0. The equation gives

$$a(E^0) = \frac{C}{E^0 - E}$$

and consequently

$$1 = V \int_{2\epsilon_F}^{2\epsilon_m} dE^{0'} \frac{\rho(E^{0'})}{E^{0'} - E}. \tag{6.262}$$

The range of interaction is small compared with $E_F^0 = 2\epsilon_F$. Therefore, $\rho(E^{0'})$ may be taken as $\rho(E_F^0)$, which can then be taken out of the integral. We get

$$\frac{1}{V\rho(E_F^0)} = \int_{2\epsilon_F}^{2\epsilon_m} \frac{dE^{0'}}{E^{0'} - E} = \ln\left(\frac{2\epsilon_m - E}{2\epsilon_F - E}\right). \tag{6.263}$$

In terms of

$$\Delta \equiv 2\epsilon_F - E,$$

this equation can be written as

$$\frac{\Delta}{2\epsilon_m - (2\epsilon_F - \Delta)} = \exp\left[\frac{-1}{V\rho(E_F^0)}\right],$$

giving

$$\Delta = 2\hbar\omega_m\{\exp[1/V\rho(E_F^0)] - 1\}^{-1}. \tag{6.264}$$

In conclusion, the state $|\psi\rangle$ considered is made up of couples of one-electron states with antiparallel spins and $\epsilon \gtrless \epsilon_F$. The energy E of $|\psi\rangle$ is less than $2\epsilon_F$ by Δ, if V is positive. With a pair of electrons in such a $|\psi\rangle$, a Cooper pair, the energy of the system is lowered by the binding energy Δ of the pair.

2. THEORY OF SUPERCONDUCTIVITY

Consider the Hamiltonian (6.244), omitting the term that does not contain electron operators. We shall for brevity let $\mathbf{k}$ stand for $\mathbf{k}, \uparrow$ and $-\mathbf{k}$ stand for $-\mathbf{k}, \downarrow$; in the abbreviated notation, $(-\mathbf{k})$ stands for $-\mathbf{k}, \uparrow$. The Hamiltonian of the one-electron approximation is then

$$H_0 = \sum_{\mathbf{k}} \epsilon_{\mathbf{k}} (c_{\mathbf{k}}^+ c_{\mathbf{k}} + c_{-\mathbf{k}}^+ c_{-\mathbf{k}}).$$

Regarding $H_{\rm I}$, the consideration about Cooper pairs indicates that the association of one-electron states with opposite wavevectors and spins lowers the energy of the system. Accordingly, the Hamiltonian to be considered for the superconducting state is, in the abbreviated notation,

$$\begin{aligned} H &= H_0 + H_{\rm I} \\ &= \sum_{\mathbf{k}} \epsilon_{\mathbf{k}} (c_{\mathbf{k}}^+ c_{\mathbf{k}} + c_{-\mathbf{k}}^+ c_{-\mathbf{k}}) + 2 \sum_{\mathbf{kq}} V_{\mathbf{k},-\mathbf{k},\mathbf{q}} c_{\mathbf{k}-\mathbf{q}}^+ c_{-(\mathbf{k}-\mathbf{q})}^+ c_{-\mathbf{k}} c_{\mathbf{k}} \\ &\equiv \sum_{\mathbf{k}} \epsilon_{\mathbf{k}} (c_{\mathbf{k}}^+ c_{\mathbf{k}} + c_{-\mathbf{k}}^+ c_{-\mathbf{k}}) + \sum_{\mathbf{kk'}} V_{\mathbf{kk'}} c_{\mathbf{k'}}^+ c_{-\mathbf{k'}}^+ c_{-\mathbf{k}} c_{\mathbf{k}}; \end{aligned} \tag{6.265}$$

obviously, $V_{\mathbf{kk}} = 0$. This is the Hamiltonian of the BCS theory. Define the following operators, sometimes called quasiparticle operators:

$$\begin{aligned} \xi_{\mathbf{k}}^+ &= u_{\mathbf{k}} c_{\mathbf{k}}^+ - v_{\mathbf{k}} c_{-\mathbf{k}}, \qquad & \xi_{\mathbf{k}} &= u_{\mathbf{k}} c_{\mathbf{k}} - v_{\mathbf{k}} c_{-\mathbf{k}}^+, \\ \xi_{-\mathbf{k}}^+ &= u_{\mathbf{k}} c_{-\mathbf{k}}^+ + v_{\mathbf{k}} c_{\mathbf{k}}, \qquad & \xi_{-\mathbf{k}} &= u_{\mathbf{k}} c_{-\mathbf{k}} + v_{\mathbf{k}} c_{\mathbf{k}}^+. \end{aligned} \tag{6.266}$$

The coefficients $u_{\mathbf{k}}$ and $v_{\mathbf{k}}$ are real and positive, and

$$u_{\mathbf{k}}^2 + v_{\mathbf{k}}^2 = 1; \tag{6.267}$$

within the limitation of these conditions, the coefficients are arbitrary. The operators have the following anticommutation relations:

$$\{\xi_{\mathbf{k'}}^+, \xi_{\mathbf{k}}\} = \delta_{\mathbf{k'k}}, \quad \{\xi_{\mathbf{k}}, \xi_{\mathbf{k'}}\} = 0, \quad \text{and} \quad \{\xi_{\mathbf{k}}^+, \xi_{\mathbf{k'}}^+\} = 0. \tag{6.268}$$

In terms of these operators, the two terms of the Hamiltonian are

$$\begin{aligned} H_0 = \sum_{\mathbf{k}} \epsilon_{\mathbf{k}} \Big[& 2v_{\mathbf{k}}^2 + (u_{\mathbf{k}}^2 - v_{\mathbf{k}}^2) (\xi_{\mathbf{k}}^+ \xi_{\mathbf{k}} + \xi_{-\mathbf{k}}^+ \xi_{-\mathbf{k}}) \\ & + 2u_{\mathbf{k}} v_{\mathbf{k}} (\xi_{\mathbf{k}}^+ \xi_{-\mathbf{k}}^+ + \xi_{-\mathbf{k}} \xi_{\mathbf{k}}) \Big] \end{aligned} \tag{6.269}$$

and

$$\begin{aligned} H_{\rm I} = \sum_{\mathbf{kk'}} V_{\mathbf{kk'}} \Big[& u_{\mathbf{k'}} u_{\mathbf{k}} v_{\mathbf{k}} v_{\mathbf{k'}} (1 - \xi_{\mathbf{k'}}^+ \xi_{\mathbf{k'}} - \xi_{-\mathbf{k'}}^+ \xi_{-\mathbf{k'}}) \\ & \times (1 - \xi_{\mathbf{k}}^+ \xi_{\mathbf{k}} - \xi_{-\mathbf{k}}^+ \xi_{-\mathbf{k}}) \\ & + u_{\mathbf{k'}} v_{\mathbf{k'}} (u_{\mathbf{k}}^2 - v_{\mathbf{k}}^2)(1 - \xi_{\mathbf{k'}}^+ \xi_{\mathbf{k'}} - \xi_{-\mathbf{k'}}^+ \xi_{-\mathbf{k'}}) \\ & \times (\xi_{-\mathbf{k}} \xi_{\mathbf{k}} + \xi_{\mathbf{k}}^+ \xi_{-\mathbf{k}}^+) + \cdots \Big]. \end{aligned} \tag{6.270}$$

For convenience of discussion, we shall measure energy from the Fermi energy of Bloch states, in which case $\epsilon_{\rm F} = 0$.

The Ground State. Consider the state $|G\rangle$ characterized by

$$\xi_{\mathbf{k}}|G\rangle = 0 = \xi_{-\mathbf{k}}|G\rangle. \tag{6.271}$$

In terms of bra vectors, we have

$$\langle G|\xi_{\mathbf{k}}^{+} = 0 = \langle G|\xi_{-\mathbf{k}}^{+}.$$

The operator N of the total number of electrons in the system is

$$\begin{aligned} N &= \sum_{\mathbf{k}}(c_{\mathbf{k}}^{+}c_{\mathbf{k}} + c_{-\mathbf{k}}^{+}c_{-\mathbf{k}}) \\ &= \sum_{\mathbf{k}}[2v_{\mathbf{k}}^2 + (u_{\mathbf{k}}^2 - v_{\mathbf{k}}^2)(\xi_{\mathbf{k}}^{+}\xi_{\mathbf{k}} + \xi_{-\mathbf{k}}^{+}\xi_{-\mathbf{k}}) + 2u_{\mathbf{k}}v_{\mathbf{k}}(\xi_{\mathbf{k}}^{+}\xi_{-\mathbf{k}}^{+} + \xi_{-\mathbf{k}}\xi_{\mathbf{k}})]. \end{aligned} \tag{6.272}$$

The state $|G\rangle$ is not an eigenstate of N, but it gives a definite expectation value for N:

$$\langle G|N|G\rangle = 2\sum_{\mathbf{k}} v_{\mathbf{k}}^2\langle G \mid G\rangle; \tag{6.273}$$

it is usually convenient to normalize $|G\rangle$ so that $\langle G \mid G\rangle = 1$. The state $|G\rangle$ under consideration is referred to as the superconducting ground state. The state vector

$$|X\rangle = \left[\prod_{\mathbf{k}}\left(u_{\mathbf{k}} + v_{\mathbf{k}}c_{\mathbf{k}}^{+}c_{-\mathbf{k}}^{+}\right)\right]|0\rangle = \left[\prod_{\mathbf{k}}\left(\frac{1}{v_{\mathbf{k}}}\right)\xi_{\mathbf{k}}\xi_{-\mathbf{k}}\right]|0\rangle, \tag{6.274}$$

where $|0\rangle$ is the vacuum state, has the characteristics of $|G\rangle$ and therefore can be taken as $|G\rangle$. Furthermore, $|X\rangle$ is normalized, $\langle X \mid X\rangle = 1$.

Let the $u_{\mathbf{k}}$ and $v_{\mathbf{k}}$ coefficients satisfy the condition

$$2\epsilon_{\mathbf{k}}u_{\mathbf{k}}v_{\mathbf{k}} + (u_{\mathbf{k}}^2 - v_{\mathbf{k}}^2)\sum_{\mathbf{k}'} V_{\mathbf{k}\mathbf{k}'}u_{\mathbf{k}'}v_{\mathbf{k}'} = 0, \tag{6.275}$$

which leads to the relations

$$u_{\mathbf{k}}v_{\mathbf{k}} = \frac{1}{2}\frac{\Delta_{\mathbf{k}}}{(\Delta_{\mathbf{k}}^2 + \epsilon_{\mathbf{k}}^2)^{1/2}}, \qquad v_{\mathbf{k}}^2 = \frac{1}{2}\left[1 - \frac{\epsilon_{\mathbf{k}}}{(\Delta_{\mathbf{k}}^2 + \epsilon_{\mathbf{k}}^2)^{1/2}}\right],$$

and

$$\Delta_{\mathbf{k}} \equiv \sum_{\mathbf{k}'} V_{\mathbf{k}\mathbf{k}'}u_{\mathbf{k}'}v_{\mathbf{k}'} = -\frac{1}{2}\sum_{\mathbf{k}'} V_{\mathbf{k}\mathbf{k}'}\frac{\Delta_{\mathbf{k}'}}{(\Delta_{\mathbf{k}'}^2 + \epsilon_{\mathbf{k}'}^2)^{1/2}}. \tag{6.276}$$

With condition (6.275), the Hamiltonian takes on the following expression:

$$H = 2\sum_{\mathbf{k}} \epsilon_{\mathbf{k}} v_{\mathbf{k}}^2 + \sum_{\mathbf{k}} \epsilon_{\mathbf{k}} (u_{\mathbf{k}}^2 - v_{\mathbf{k}}^2)(\xi_{\mathbf{k}}^+ \xi_{\mathbf{k}} + \xi_{-\mathbf{k}}^+ \xi_{-\mathbf{k}})$$
$$+ \sum_{\mathbf{k}\mathbf{k}'} V_{\mathbf{k}\mathbf{k}'} u_{\mathbf{k}'} u_{\mathbf{k}} v_{\mathbf{k}} v_{\mathbf{k}'} (1 - \xi_{\mathbf{k}'}^+ \xi_{\mathbf{k}'} - \xi_{-\mathbf{k}'}^+ \xi_{-\mathbf{k}'})(1 - \xi_{\mathbf{k}}^+ \xi_{\mathbf{k}} - \xi_{-\mathbf{k}}^+ \xi_{-\mathbf{k}})$$
$$- \sum_{\mathbf{k}\mathbf{k}'} V_{\mathbf{k}\mathbf{k}'} u_{\mathbf{k}'} v_{\mathbf{k}'} (u_{\mathbf{k}}^2 - v_{\mathbf{k}}^2)(\xi_{\mathbf{k}'}^+ \xi_{\mathbf{k}'} + \xi_{-\mathbf{k}'}^+ \xi_{-\mathbf{k}'})(\xi_{-\mathbf{k}} \xi_{\mathbf{k}} + \xi_{\mathbf{k}}^+ \xi_{-\mathbf{k}}^+).$$

(6.277)

The state $|G\rangle$ is then an eigenstate of the Hamiltonian, with energy

$$E_G = 2\sum_{\mathbf{k}} \epsilon_{\mathbf{k}} v_{\mathbf{k}}^2 + \sum_{\mathbf{k}\mathbf{k}'} V_{\mathbf{k}\mathbf{k}'} u_{\mathbf{k}} v_{\mathbf{k}} u_{\mathbf{k}'} v_{\mathbf{k}'} = 2\sum_{\mathbf{k}} \epsilon_{\mathbf{k}} v_{\mathbf{k}}^2 - \sum_{\mathbf{k}} \Delta_{\mathbf{k}} u_{\mathbf{k}} v_{\mathbf{k}}.$$

(6.278)

In the absence of interaction H_{I}, the ground-state energy E_0 given by H_0 is related to the Fermi energy ϵ_{F} in the familiar way. At sufficiently low temperatures, all states $\mathbf{k}$ and $-\mathbf{k}$ below ϵ_{F} are practically filled with electrons and those above ϵ_{F} are practically unoccupied, with

$$E_0 = 2 \sum_{|\mathbf{k}|<|\mathbf{k}_F|} \xi_{\mathbf{k}}. \tag{6.279}$$

The interaction changes the energy of the system by

$$\delta E_G \equiv E_G - E_0 = 2 \sum_{|\mathbf{k}|<|\mathbf{k}_F|} \epsilon_{\mathbf{k}} (v_{\mathbf{k}}^2 - 1) + 2 \sum_{|\mathbf{k}|>|\mathbf{k}_F|} \epsilon_{\mathbf{k}} v_{\mathbf{k}}^2 - \sum_{\mathbf{k}} \Delta_{\mathbf{k}} u_{\mathbf{k}} v_{\mathbf{k}}$$
$$= - \sum_{|\mathbf{k}|<|\mathbf{k}_F|} \epsilon_{\mathbf{k}} \left[1 + \frac{\epsilon_{\mathbf{k}}}{(\Delta_{\mathbf{k}}^2 + \epsilon_{\mathbf{k}}^2)^{1/2}}\right]$$
$$+ \sum_{|\mathbf{k}|>|\mathbf{k}_F|} \epsilon_{\mathbf{k}} \left[1 - \frac{\epsilon_{\mathbf{k}}}{(\Delta_{\mathbf{k}}^2 + \epsilon_{\mathbf{k}}^2)^{1/2}}\right] - \frac{1}{2} \sum_{\mathbf{k}} \frac{\Delta_{\mathbf{k}}^2}{(\Delta_{\mathbf{k}}^2 + \epsilon_{\mathbf{k}}^2)^{1/2}}. \tag{6.280}$$

As pointed out in connection with (6.252), only a small range around ϵ_{F} is important for the interaction. Using the approximations

$$\epsilon_{(\mathbf{k}_F - \mathbf{k})} = -\epsilon_{(\mathbf{k}_F + \mathbf{k})} \quad \text{and} \quad \Delta_{(\mathbf{k}_F - \mathbf{k})}^2 = \Delta_{(\mathbf{k}_F + \mathbf{k})}^2, \tag{6.281}$$

we get

$$\delta E_G = 2 \sum_{|\mathbf{k}|>|\mathbf{k}_F|} \left[\epsilon_{\mathbf{k}} - \frac{2\epsilon_{\mathbf{k}}^2 + \Delta_{\mathbf{k}}^2}{2(\epsilon_{\mathbf{k}}^2 + \Delta_{\mathbf{k}}^2)^{1/2}}\right]. \tag{6.282}$$

It will be assumed that $V_{\mathbf{k}\mathbf{k}'}$ has a constant negative value within a small range of ϵ around ϵ_F and that it is zero outside this range:

$$V_{\mathbf{k}\mathbf{k}'} = \begin{cases} -V & \text{for } |\epsilon_{\mathbf{k}}|, |\epsilon_{\mathbf{k}'}| < \hbar\omega_m \\ 0 & \text{otherwise.} \end{cases} \tag{6.283}$$

Consequently,

$$\Delta_{\mathbf{k}} = \begin{cases} \Delta & \text{for } |\epsilon_{\mathbf{k}}| < \hbar\omega_m \\ 0 & \text{otherwise,} \end{cases} \tag{6.284}$$

in which Δ is given by the equation

$$1 = \tfrac{1}{2}V \sum_{\mathbf{k}} (\Delta^2 + \epsilon_{\mathbf{k}}^2)^{-1/2}. \tag{6.285}$$

Let ρ_ϵ denote the density of states of one spin at ϵ. The equation for Δ can be written as

$$1 = \frac{1}{2}V \int_{-\hbar\omega_m}^{\hbar\omega_m} d\epsilon \frac{\rho_\epsilon}{(\Delta^2 + \epsilon^2)^{1/2}} \simeq \frac{1}{2}V \rho_F \int_{-\hbar\omega_m}^{\hbar\omega_m} \frac{d\epsilon}{(\Delta^2 + \epsilon^2)^{1/2}}, \tag{6.286}$$

giving

$$\Delta = \frac{\hbar\omega_m}{\sinh(1/V\rho_F)} \approx 2\hbar\omega_m \exp\left(\frac{-1}{V\rho_F}\right). \tag{6.287}$$

With $\Delta_{\mathbf{k}} = \Delta$ and the summation replaced by integration, the equation of δE_G becomes

$$\begin{aligned} \delta E_G &= 2\rho_F \int_0^{\hbar\omega_m} \left[\epsilon - \frac{2\epsilon^2 + \Delta^2}{2(\epsilon^2 + \Delta^2)^{1/2}}\right] d\epsilon \\ &= (\hbar\omega_m)^2 \rho_F [1 - \coth(1/V\rho_F)] \approx -\tfrac{1}{2}\rho_F \Delta^2. \end{aligned} \tag{6.288}$$

We see that δE_G is negative, that is, the superconducting ground state $|G\rangle$ has an energy lower than the ground state energy E_0 in the absence of interaction H_I.

The adopted relation (6.275) between $u_{\mathbf{k}}$ and $v_{\mathbf{k}}$ makes the state $|G\rangle$ given by (6.274) an eigenstate of the Hamiltonian. Consider now the expectation value $\langle G|N|G\rangle = 2\sum_{\mathbf{k}} v_{\mathbf{k}}^2$ of the total number of electrons.

$$\langle G|N|G\rangle = 2\sum_{\mathbf{k}} v_{\mathbf{k}}^2 = \sum_{\mathbf{k}} \left(1 - \frac{\epsilon_{\mathbf{k}}}{(\epsilon_{\mathbf{k}}^2 + \Delta_{\mathbf{k}}^2)^{1/2}}\right). \tag{6.289}$$

It follows that

$$\langle G|N|G\rangle = 2 \sum_{|\mathbf{k}|<|\mathbf{k}_F|} 1 \tag{6.290}$$

in cases where the approximations (6.281) apply. Physically, this result is expected under such approximations.

Excited States. Consider the state

$$|\mathbf{k}_0\rangle = \xi^+_{-\mathbf{k}_0}\xi^+_{\mathbf{k}_0}|G\rangle$$
$$= \left(u_{\mathbf{k}_0}c^+_{-\mathbf{k}_0} + v_{\mathbf{k}_0}c_{\mathbf{k}_0}\right)c^+_{\mathbf{k}_0}\prod_{\mathbf{k}\neq\mathbf{k}_0}\left(u_\mathbf{k} + v_\mathbf{k}c^+_\mathbf{k}c^+_{-\mathbf{k}}\right)|0\rangle. \tag{6.291}$$

To evaluate the energy $E_{\mathbf{k}_0}$ of the state, disregard the last sum of terms in expression (6.277), since it has zero expectation value. Furthermore, omit terms in the third sum that are of fourth order in ξ; they are negligible at $T = 0$. We get

$$E_{\mathbf{k}_0} = E_G + 2(u^2_{\mathbf{k}_0} - v^2_{\mathbf{k}_0})\epsilon_{\mathbf{k}_0} - 4u_{\mathbf{k}_0}v_{\mathbf{k}_0}\sum_{\mathbf{k}\neq\mathbf{k}_0} V_{\mathbf{k}_0\mathbf{k}}u_\mathbf{k}v_\mathbf{k}. \tag{6.292}$$

In view of (6.276), we have

$$\mathcal{E}_{\mathbf{k}_0} \equiv E_{\mathbf{k}_0} - E_G = 2(\epsilon^2_{\mathbf{k}_0} + \Delta^2_{\mathbf{k}_0})^{1/2}, \tag{6.293}$$

which is the excitation energy of state $|\mathbf{k}_0\rangle$.

We note that $|\mathbf{k}_0\rangle$ is a state of the system with

$$\langle\mathbf{k}_0|N|\mathbf{k}_0\rangle = \sum_\mathbf{k} 2v^2_\mathbf{k} + 2(u^2_{\mathbf{k}_0} - v^2_{\mathbf{k}_0}), \tag{6.294}$$

whereas $|G\rangle$ is the ground state of a system with

$$\langle G|N|G\rangle = \sum_\mathbf{k} 2v^2_\mathbf{k}.$$

From this point of view, the state expressed by (6.291) and its excitation energy given by (6.293) strictly apply only for the case of

$$u^2_{\mathbf{k}_0} = v^2_{\mathbf{k}_0},$$

which requires $\epsilon_{\mathbf{k}_0} = 0$, that is, $\epsilon_{\mathbf{k}_0}$ is the Fermi level. However, consider an excited state that behaves as an average of two parts involving, respectively, $\mathbf{k}_0$ and $\mathbf{k}'_0$. With $\epsilon_{\mathbf{k}_0} = -\epsilon_{\mathbf{k}_0'}$

$$u^2_{\mathbf{k}_0} - v^2_{\mathbf{k}_0} = -(u^2_{\mathbf{k}_0'} - v^2_{\mathbf{k}_0'}) \qquad \text{and} \qquad \epsilon^2_{\mathbf{k}_0} + \Delta^2_{\mathbf{k}_0} = \epsilon^2_{\mathbf{k}_0'} + \epsilon^2_{\mathbf{k}'}$$

For such an excited state, the excitation energy is given by $2(\epsilon_{\mathbf{k}_0} + \Delta_{\mathbf{k}_0})$ without the limitation $\epsilon_{\mathbf{k}_0} = 0$. The quantity $\Delta_{\mathbf{k}_0}$, or simply Δ under assumption (6.284), gives the minimum excitation energy 2Δ and is called the gap parameter. The gap is fundamental for the phenomena of superconductivity. Equation (6.270) shows that the $V_{\mathbf{kk}'}$'s produce the interaction Hamiltonian H_{I}, and (6.276) shows directly that Δ comes from the existence of $V_{\mathbf{kk}'}$'s.

For convenience, an excited state considered here is often called a quasiparticle state.

Transition Temperature. The energy of an excitation is affected by the presence of other excitations. This effect has so far been left out of consideration by neglecting $(\xi_{\mathbf{k}'}\xi_{\mathbf{k}'}+\xi_{-\mathbf{k}'}\xi_{-\mathbf{k}'})(\xi_{\mathbf{k}}\xi_{\mathbf{k}}+\xi_{-\mathbf{k}}\xi_{-\mathbf{k}})$ in the third sum of (6.277). It is correct to do so for temperature $T=0$. According to the Fermi–Dirac distribution, the population of a quasiparticle state is given by

$$f[\mathcal{E}_{\mathbf{k}}(T)]=\left[\exp\left(\frac{\mathcal{E}_{\mathbf{k}}(T)}{kT}\right)+1\right]^{-1}. \tag{6.295}$$

For extension to finite temperatures,

$$\mathcal{E}_{\mathbf{k}}(T)=2(u_{\mathbf{k}}^2-v_{\mathbf{k}}^2)_{\mathbf{k}}-4u_{\mathbf{k}}v_{\mathbf{k}}\sum_{\mathbf{k}'\neq\mathbf{k}}V_{\mathbf{kk}'}u_{\mathbf{k}'}v_{\mathbf{k}'}[1-f_{\mathbf{k}'}(T)-f_{-\mathbf{k}'}(T)] \tag{6.296}$$

may be taken in place of (6.292). The factor

$$[1-f_{\mathbf{k}'}(T)-f_{-\mathbf{k}'}(T)]=\tanh\left[\frac{\mathcal{E}_{\mathbf{k}'}(T)}{2kT}\right]$$

has the apparent effect of modifying $V_{\mathbf{kk}'}$. Consequently, (6.275) is changed to

$$2\epsilon_{\mathbf{k}}u_{\mathbf{k}}v_{\mathbf{k}}+(u_{\mathbf{k}}^2-v_{\mathbf{k}}^2)\sum_{\mathbf{k}'}V_{\mathbf{kk}'}\tanh\left[\frac{\mathcal{E}_{\mathbf{k}'}(T)}{2kT}\right]u_{\mathbf{k}'}v_{\mathbf{k}'}=0, \tag{6.297}$$

and (6.276) is changed to

$$\Delta_{\mathbf{k}}(T)=-\frac{1}{2}\sum_{\mathbf{k}'}V_{\mathbf{kk}'}\tanh\left[\frac{\mathcal{E}_{\mathbf{k}'}(T)}{2kT}\right]\Delta_{\mathbf{k}'}(T)[\Delta_{\mathbf{k}'}^2(T)+\epsilon_{\mathbf{k}'}^2]^{-1/2}. \tag{6.298}$$

The excitation energy is given by

$$\mathcal{E}_{\mathbf{k}}(T)=2[\epsilon_{\mathbf{k}}^2+\Delta_{\mathbf{k}}^2(T)]^{1/2}, \tag{6.299}$$

which involves the temperature-dependent gap parameter $\Delta_{\mathbf{k}}(T)$.

Solution of (6.298) gives a $\Delta_{\mathbf{k}}(T)$ that decreases with increasing T. The temperature T_c at which $\Delta_{\mathbf{k}}(T)=0$ is called the transition temperature. Using the assumptions leading to (6.286), we have at T_c

$$1 = V\rho_F \int_0^{\hbar\omega_m} \frac{\tanh(\epsilon/2kT_c)}{\epsilon}\, d\epsilon. \tag{6.300}$$

Usually, $kT_c \ll \hbar\omega_m$. We get with good approximation

$$kT_c = 1.13\hbar\omega_m \exp(-1/V\rho_F) = 2\Delta/3.53. \tag{6.301}$$

The Debye frequency ω_m is inversely proportional to the square root of the atomic mass M. Equation (6.301) thus shows that $T_c \propto M^{-1/2}$, which is called the isotope effect.

3. PHENOMENA OF SUPERCONDUCTIVITY

a. Persistent Current

Persistent current or zero electrical resistance is the basic property of superconductivity. Electrical resistance is produced by the scattering of the drifting electrons. The absence of electrical resistance in a superconductor can be understood from the following simple consideration.

In superconductivity, electrons of $\mathbf{k}$ and $-\mathbf{k}$ ($\mathbf{k},\uparrow$ and $-\mathbf{k},\downarrow$) are paired by virtue of electron–electron interaction involving phonons. We can instead pair electrons of $\mathbf{k}+\mathbf{k}_d$ and $-\mathbf{k}+\mathbf{k}_d$ ($\mathbf{k}+\mathbf{k}_d,\uparrow$ and $-\mathbf{k}+\mathbf{k}_d,\downarrow$) and consequently obtain a state that is entirely equivalent to the original superconducting state viewed from a coordinate system moving with a velocity $-\hbar\mathbf{k}_d/m$. In other words, we get an equivalent superconducting state for electrons drifting with a velocity $\hbar\mathbf{k}_d/m$; the electric current density is $Ne\hbar\mathbf{k}_d/m$, N being the concentration of electrons. For electrons with a steady drift velocity $\mathbf{v}_d \equiv \hbar\mathbf{k}_d/m$, the Fermi surface in the wavenumber space is shifted by $\mathbf{k}_d$. The scattering of electrons from the leading edge of the Fermi surface to the trailing edge decreases the drift current, giving rise to electrical resistance. In a normal conductor, such scattering requires from the energy point view only that $\hbar^2k^2/2m - \hbar^2k'^2/2m$ be equal to zero or equal to the energy expenditure of the scattering process. In a superconductor, on the other hand, $\hbar^2k^2/2m - \hbar^2k'^2/2m$ must also cover the 2Δ required for quasiparticle excitation. Therefore, scattering can occur only if $(\hbar^2/m)2k_Fk_d \geq 2\Delta$, that is, only if the electric current exceeds $Ne\Delta/\hbar k_F$ in magnitude. A current within the limit is not decreased by scattering.

b. The Meissner Effect

An important effect in superconductivity, the Meissner effect, was discovered in 1933. It consists of the expulsion of an applied magnetic field by a superconductor. The effect is explained in the following.

One-electron states in the presence of a steady magnetic field have been treated in Section IV.A.3. Consider the Hamiltonian of a system of electrons, $H = H_0 + H_A$, where H_0 is independent of the field and H_A is linear in the vector potential $\mathbf{A}$ of the field. The spin contribution and the contribution of higher order in $\mathbf{A}$ will not be considered. In the second quantization form, we have

$$H_A = -\frac{e}{m^*c}\int d\mathbf{r}\,\Psi^+(\mathbf{r})\mathbf{A}\cdot\mathbf{p}\Psi(r). \qquad (6.302)$$

The effective mass m^* of electrons has been taken to be scalar. Using the following approximation for the field operators $\Psi(\mathbf{r})$ and $\Psi^+(\mathbf{r})$,

$$\Psi(\mathbf{r}) = \sum_k c_{\mathbf{k}}\varphi_{\mathbf{k}}(\mathbf{r}) \sim \sum_{\mathbf{k}} c_{\mathbf{k}} \exp(i\mathbf{k}\cdot\mathbf{r})$$

and

$$\Psi^+(\mathbf{r}) \sim \sum_{\mathbf{k}} c_{\mathbf{k}}^+ \exp(-i\mathbf{k}\cdot\mathbf{r}), \qquad (6.303)$$

we get

$$H_A = -\frac{e}{m^*c}\sum_{\mathbf{k}\kappa}\int d\mathbf{r}\, c_{\mathbf{k}+\kappa}^+ c_{\mathbf{k}}\mathbf{A}\cdot\mathbf{k}\exp(-i\kappa\cdot\mathbf{r}). \qquad (6.304)$$

For a vector potential

$$\mathbf{A}(\mathbf{r}) = \mathbf{a}_{\mathbf{q}}\exp(i\mathbf{q}\cdot\mathbf{r}), \qquad (6.305)$$

we have

$$H_A = -\frac{e}{m^*c}\sum_{\mathbf{k}} c_{\mathbf{k}+\mathbf{q}}^+ c_{\mathbf{k}}\mathbf{k}\cdot\mathbf{a}_{\mathbf{q}}. \qquad (6.306)$$

Consider now the current density of electrons, the operator $\mathbf{J}$ of which is given by

$$\begin{aligned}\mathbf{J}(\mathbf{r}) &= \frac{e\hbar}{2m^*i}(\Psi^+\nabla\Psi - \Psi\nabla\Psi^+) - \frac{e^2}{m^*c}\Psi^+\mathbf{A}\Psi \\ &= \frac{e\hbar}{2m^*}\sum_{\mathbf{k}\boldsymbol{\kappa}} c^+_{\mathbf{k}+\boldsymbol{\kappa}}c_{\mathbf{k}}\exp(-i\boldsymbol{\kappa}\cdot\mathbf{r})(2\mathbf{k}+\boldsymbol{\kappa}) \\ &\quad - \frac{e^2}{m^*c}\sum_{\mathbf{k}\boldsymbol{\kappa}} c^+_{\mathbf{k}+\boldsymbol{\kappa}}c_{\mathbf{k}}\exp(-i\boldsymbol{\kappa}\cdot\mathbf{r})\mathbf{A}(\mathbf{r}) \\ &\equiv \mathbf{J}_1(\mathbf{r}) + \mathbf{J}_2(\mathbf{r}). \end{aligned} \tag{6.307}$$

Following perturbation theory, we write the state of the electron system as

$$S = S_0 + S' + \cdots, \tag{6.308}$$

where S_0 is independent of $\mathbf{A}$ and S' is linear in $\mathbf{A}$. Let $|0\rangle$ be the particular state and $|m\rangle$'s be the other states given by H_0. We have

$$S' = \sum_m{}' |m\rangle \frac{\langle m|H_A|0\rangle}{E_0 - E_m}. \tag{6.309}$$

The component $\mathbf{J}_2$ is linear in $\mathbf{A}$. To the first order of $\mathbf{A}$, its expectation value is

$$\mathbf{j}_2 = \langle S_0|\mathbf{J}_2|S_0\rangle = -\frac{e^2}{m^*c}\langle S_0|\sum_{\mathbf{k}\boldsymbol{\kappa}} c^+_{\mathbf{k}+\boldsymbol{\kappa}}c_{\mathbf{k}}\exp(-i\boldsymbol{\kappa}\cdot\mathbf{r})\mathbf{A}(\mathbf{r})|S_0\rangle.$$

We are interested in vector potentials in the limit of $q \to 0$. In this case,

$$\mathbf{j}_2(\mathbf{r}) = -\frac{Ne^2}{m^*c}\mathbf{A}(\mathbf{r}), \tag{6.310}$$

where N is the density of electrons. With regard to the magnetic susceptibility given by (6.176), $\mathbf{j}_2$ is the current component that gives diamagnetism.

The component of current, $\mathbf{j}_1$, gives a paramagnetic contribution in addition to that associated with spin. With regard to $\mathbf{j}_1$, the superconducting state differs essentially from the normal states. With $\mathbf{J}_1$ independent of $\mathbf{A}$, $\langle S_0|\mathbf{J}_1|S_0\rangle = 0$. To the first order of $\mathbf{A}$, the expectation value of $\mathbf{J}_1$ is

$$\mathbf{j}_1 = \langle S_0|\mathbf{J}_1|S'\rangle + \langle S'|\mathbf{J}_1|S_0\rangle, \tag{6.311}$$

where

$$\langle S_0|\mathbf{J}_1|S'\rangle = \sum_m{}' \frac{1}{E_0 - E_m}\langle 0|\mathbf{J}_1|m\rangle\langle m|H_A|0\rangle. \tag{6.312}$$

For a superconductor, consider S_0 to be the superconducting ground state $|G\rangle$, and take the $|m\rangle$'s to be the excited states:

$$|m\rangle = \xi^{+}_{-\kappa_m+\mathbf{q}}\xi^{+}_{\mathbf{k}_m}|G\rangle. \tag{6.313}$$

It can be shown that as $q \to 0$,

$$\langle m|H_A|G\rangle \to 0 \qquad \text{and} \qquad (E_G - E_m) \to 2(\epsilon_{\mathbf{k}_m} + \Delta).$$

It follows that

$$\mathbf{j}_1 = \langle G|\mathbf{J}_1|S'\rangle \to 0, \tag{6.314}$$

a result that is true only for a superconductor. This result leads to

$$\mathbf{j}(\mathbf{r}) = \mathbf{j}_2(\mathbf{r}) = -\frac{Ne^2}{m^*c}\mathbf{A}(\mathbf{r}). \tag{6.315}$$

This is known as the London equation and was postulated before the establishment of the theory of superconductivity.

We have for a static magnetic field the Maxwell equation

$$\nabla \times \mathbf{H} = \frac{1}{\mu}\nabla \times \nabla \times \mathbf{A} = \frac{4\pi}{c}\mathbf{j}, \tag{6.316}$$

μ being the magnetic permeability. When the London equation applies, we get

$$\nabla^2\mathbf{j} = \frac{1}{2}\mathbf{j} \qquad \text{and} \qquad \nabla^2\mathbf{B} = \frac{1}{\lambda^2}\mathbf{B}, \tag{6.317}$$

where

$$\lambda = (m^*c^2/4\pi Ne^2\mu)^{1/2}. \tag{6.318}$$

For a magnetic field just outside and parallel to the surface of the superconductor, at $z = 0$, the field inside the superconductor is given by

$$\mathbf{B}(z) = \mathbf{B}_0 e^{-z/\lambda}. \tag{6.319}$$

The parameter λ is known as the penetration depth and is of the order of 10^{-15} cm. A normal conductor in a magnetic field expels the field as it becomes superconducting upon cooling. This is the Meissner effect. The parameter λ is apparently a function of temperature, becoming infinite when the conductor becomes perfectly normal. Expression (6.318), obtained by taking $|G\rangle$ as S_0, gives $\lambda(T = 0)$.

A consequence of the Meissner effect is the destruction of superconductivity by a sufficiently high magnetic field. The repulsion of magnetic field requires an energy of $B^2/8\pi$ per unit volume. This energy has to be covered by δE if the material is to be superconducting, where δE is the reduction of energy density in comparison with that of the normal state. For low temperatures, $T \sim 0$, δE is δE_G given by (6.288), and the critical field B_c above which the material cannot be superconducting is given by

$$B_c/8\pi = \delta E_G. \tag{6.320}$$

Flux Quantization. A remarkable phenomenon of superconductivity is the quantization of magnetic flux that goes through a closed ring of superconductor. Inside the superconductor, deeper than the small penetration depth, $\nabla \times \mathbf{A} = \mathbf{B} = 0$ according to the Meissner effect. The vector potential can be expressed as the gradient of a scalar $\chi(\mathbf{r})$:

$$\mathbf{A} = \nabla\chi. \tag{6.321}$$

Going around the ring, we find χ to have changed by

$$\delta\chi = \oint \nabla\chi \cdot d\mathbf{l} = \oint \mathbf{A} \cdot d\mathbf{l} = \Phi \tag{6.322}$$

each time a closed loop is completed; Φ is the magnetic flux through the ring and a small penetration depth in the superconductor. The equation shows that χ cannot be a constant and that it is not single-valued.

Consider the wave equation for the system of electrons in the ring:

$$\left\{\sum_i \frac{1}{2m}\left[\mathbf{p}_i - \frac{e}{c}\mathbf{A}(\mathbf{r}_i)\right]^2 + V\right\}\psi = E\psi. \tag{6.323}$$

Introduce the function

$$\psi' = \psi \exp\left[\sum_i \frac{-ie}{c\hbar}\chi(\mathbf{r}_i)\right]. \tag{6.324}$$

In view of (6.321), we get

$$\left\{\frac{1}{2m}\sum_i \mathbf{p}_i^2 + V\right\}\psi' = E\psi'. \tag{6.325}$$

The equation has the same form as that for ψ with $\mathbf{A} = 0$. As a solution of this equation, ψ' is single-valued. In the superconducting state, electrons are paired; a pair has a charge $2e$. According to its definition, ψ' is single-valued if the change $\delta\chi(\mathbf{r})$ for complete loops around the ring conforms to

$$(2e/c\hbar)\delta\chi = \text{integer} \times 2\pi.$$

In view of (6.322), we have

$$\Phi = \text{integer} \times (2\pi\hbar c/2e), \tag{6.326}$$

that is, the flux Φ through the superconducting ring is quantized.

c. The Coherence Length

The coherence length is a fundamental parameter in superconductivity. Superconductivity pertains to pairs of electrons. The two electrons of a pair form a wave packet. For electrons in an energy range of Δ around the Fermi energy, the linear spatial dimension of a packet is of the order of $|\partial\epsilon/\partial k|_{\mathrm{F}}/\Delta = \hbar v_{\mathrm{F}}/\Delta$. A space of such a linear dimension usually contains a huge number of pairs, since the dimension is much larger than the linear separation of electrons for a normal density of electrons. The qualitative consideration shows that the dimension discussed is a measure of the degree to which a system of pairs forms a coherent state. In theoretical treatments, a coherence length ξ is defined such that

$$\xi = \hbar v_{\mathrm{F}}/\pi\Delta. \tag{6.327}$$

Other definitions have been made that differ from this by a numerical factor.

The distance of coherence must be considered for all properties of a superconductor. According to the London equation, (6.315), applied for the Meissner effect, the current density at a point $\mathbf{r}$, $\mathbf{j}(\mathbf{r})$, is determined by the vector potential at the point, $A(\mathbf{r})$. As pointed out, the London equation applies for $q \to 0$. In general, $\mathbf{j}(\mathbf{r})$ of a system of coherent wavepackets depends on $\mathbf{A}(\mathbf{r}+\delta\mathbf{r})$ in a region surround the point $\mathbf{r}$. The effectiveness of $\mathbf{A}(\mathbf{r}+\delta\mathbf{r})$ decreases with increasing δr, becoming negligible for $\delta r \gg \xi$. The London equation must be modified if $\mathbf{A}$ has a spatial variation that is not slow in terms of ξ.

A theory of superconductivity was developed phenomenologically by Ginsberg and Landau prior to the appearance of miscroscopic BCS theory. It expresses the fraction $n_s(\mathbf{r})$ of electrons that are superconducting through an order parameter $\Phi(\mathbf{r}) : n_s(\mathbf{r}) = \Phi^*(\mathbf{r})\Phi(\mathbf{r})$. The free energy $f(\mathbf{r})$ at $\mathbf{r}$ is expressed as an expansion in powers of $n_s(\mathbf{r})$; only the two terms n_s and n_s^2 were considered, the theory being mainly for small n_s at temperatures not much lower than T_c. A term was added to take into account any spatial variation due to inhomogeneity of the material and the existing vector potential $\mathbf{A}(\mathbf{r})$. The free energy of the materials is $F = \int f(\mathbf{r})\, d\mathbf{r}$. The condition for equilibrium of the material is $\delta F = 0$, where δF is the variation of F with respect to Φ, Φ^*, and $\mathbf{A}$. This condition gives an equation for determining the order parameter Φ to be used to calculate the macroscopic properties of the material. The Ginsberg–Landau theory is convenient for applications, and it has since been derived by L. P. Gorkov from the microscopic theory.

The Ginsberg–Landau theory involves a characteristic parameter κ known as the Landau–Ginsberg parameter. It has been shown that this dimensionless parameter is the ratio of the penetration depth to the coherence length:

$$\kappa = \lambda/\xi. \tag{6.328}$$

Superconductors having $\kappa < 1$ are called type I superconductors, and those with $\kappa > 1$ are type II superconductors. A superconductor of type I has one critical field H_c, below which it shows the full Meissner effect and above which superconductivity vanishes. In a superconductor of type II, there are two critical fields, between which there is an intermediate state with superconducting and normal regions mixed. Simple and reasonably pure metals are likely to be type I superconductors. Impure metals and transition metals are usually type II superconductors.

d. Tunneling

Consider the tunneling of charge carriers through a thin insulating layer between two metals, at least one of which is a superconductor. The phenomenon is important for practical applications and is useful for experimental studies of the theory of superconductivity. The effect of tunneling of charge carriers as individual electrons was observed by I. Giaever. Subsequently, the effect of electron-pair tunneling was revealed in the work of B. D. Josephson.

For simplicity, consider $T = 0$, and consider first the Giaever tunneling. The two metals in question have the same chemical potential, Fermi energy ϵ_F, at equilibrium. In a normal metal, one-electron states with energies below ϵ_F are filled and those with energies above ϵ_F are empty of electrons. In a superconductor, on the other hand, the filled states have energies below $\epsilon_F - \Delta$, and the empty states have energies above $\epsilon_F + \Delta$. If one of the metals is normal, the flow of electrons can occur only when the applied voltage V exceeds Δ/e.

If both materials are superconductors, the flow of electrons starts at $V = (\Delta_1 + \Delta_2)/e$.

For a junction of two superconductors, tunneling of electron pairs (Josephson tunneling) comes into consideration. We give here a sketchy discussion of its effects by using the order parameter. Let the order parameters of the two superconductors be written as $P_1(\mathbf{r})e^{i\varphi_1}$ and $P_2(r)e^{i\varphi_2}$, respectively, with real P's and phases φ_1 and φ_2. The electron wavefunctions and therefore the order parameters of the two superconductors overlap. It can be shown that the supercurrent from 1 to 2 due the flow of electron pairs is

$$J = -C\sin(\varphi_2 - \varphi_1) \equiv -C\sin\varphi, \qquad (6.329)$$

where C is a positive constant characterizing the overlap, the current being opposite in sign to the flow of electron pairs. In the absence of a potential difference between the two superconductors, a dc current flows across the junction if the inherent phases of the superconductors are different; for small phase differences, current flows into the superconductor with smaller phase.

It can be shown that a potential difference between the two superconductors, $V = V_2 - V_1$, produces a variation of φ with time:

$$\hbar\frac{d\varphi}{dt} = -2eV. \qquad (6.330)$$

The effect can be roughly understood on the basis that a shift $e\delta V$ of the potential energy of each electron produces a phase change of $-(2e\delta V)t/\hbar$. The two equations (6.329) and (6.330) together show that a dc voltage V across the junction gives an ac current. This is a remarkable property of Josephson tunneling. Another interesting property of Josephson tunneling is the dependence of current on the magnetic flux present in the junction, given by

$$J \propto \frac{\sin[\pi(\phi/\phi_0)]}{\pi(\phi/\phi_0)}, \qquad (6.331)$$

where ϕ is the flux in the junction and $\phi_0 = hc/2e$ is the flux quantum.

F. STRUCTURE

The structure of matter of specified composition is a very difficult problem for theoretical treatment. It depends on external parameters such as temperature, applied forces, and applied fields of various kinds. Of basic importance is the cohesive energy, the energy required to separate the matter at $T = 0$ into isolated atoms or molecules. Evidently, a theoretical treatment of cohesive energy accurately taking into account all interactions of electrons and ions is not practicable. Various approximations have been made that are useful to some extent.

Crystalline solids are usually classified in terms of the dominant type of binding.

Molecular crystals. A so-called molecular crystal consists of electrically neutral atoms or molecules that are bound together by van der Waals forces. The atoms or molecules are nonpolar, having no average electric moment. The electrostatic van der Waals force arises from the fluctuating electric moment of individual atoms or molecules given by the motion of their electrons. The binding is apparently quite weak. The inert gas crystals are typical crystals of this type.

Ionic crystals An ionic crystal is supposed to consist of positive and negative ions. There may be more than one kind of ion of each type, but the total system is electrically neutral. The alkali halides, CuO_2, and Al_2O_3 are examples of what may be considered crystals of this category. The interaction energy of nonoverlapping ions each of which has a spherically symmetrical charge distribution is given by

$$E_e = \frac{1}{2} \sum_{i,j \neq i} \frac{q_i q_j}{r_{ij}},$$

where i, j are ion indices, q is the charge of the ion, and r is the distance between the centers of two ions. For two-atom crystals, a proportionality factor, the Madelung constant, that is a characteristic parameter of the lattice has been worked out. The energy E_e represents the main attractive interaction for an ionic crystal. For better accuracy, the van der Waals interaction can also be considered.

Mutual repulsion of the ions is necessary to stabilize the crystal and prevent it from collapse. Obviously, the repulsion comes from the overlapping of electrons on different ions, which is theoretically an intricate problem. We will not discuss here the various approximations that have been used.

Covalent crystals. A covalent bond is the binding of two atoms by a pair of electrons, one from each atom. It has been the subject of extensive treatment in molecular physics. In a so-called covalent crystal, two neighboring atoms may, to the first order of approximation, be considered as bound by a covalent bond. The monatomic crystals of a group IV element—diamond, silicon, germanium, and gray tin—are a simple prototype, each atom having equivalent bonds with all its neighbors. Each atom having four valence electrons, the lattice structure has to be of the diamond type, which is characterized by four neighbors for each atom arranged with tetrahedral symmetry.

Many diatomic compounds have a lattice structure of the zinc blende type that is similar to the diamond type in that each atom has four neighbors located with tetrahedral symmetry. With two kinds of atoms differing in the number of valence electrons, the simple consideration of covalent bond is not adequate. Some treatments of such compounds have used a mixture of covalent and ionic bonding as an approximation.

Metals. In a crystal of this category, important for the cohesion are the electrons which have to be considered as belonging to the crystal rather than to individual atoms or molecules. For a given type of lattice structure, the crystal energy, which can be calculated following the discussions in Chapter III, is a function of the lattice dimension. For large dimensions, it is simply the sum of energies of isolated atoms. As the wavefunctions of an electron-occupied state of individual atoms begin to overlap with decreasing lattice dimension, the state becomes a one-electron energy band of the crystal. The energy band broadens with decreasing lattice dimension, and its average energy first decreases, passes a minimum, and then rises sharply. The energy obtained by taking into account the electrons in all energy bands has a minimum that corresponds to the actual lattice dimension. The difference between the energy for large lattice dimensions and the energy minimum is the cohesive energy.

G. PHASE TRANSITION

For a substance of given composition, the equilibrium state is determined by a sufficient set of parameters pertaining to its contact with the external medium. For example, two of the three "external" parameters, temperature, pressure, and specific volume, may constitute a sufficient set for a simple case. Any macroscopic observable of the substance in equilibrium has a functional relationship with the set of parameters. We will consider this relationship as the equation of state for the particular observable.

A phase transition with respect to an observable refers to a discontinuity or a singularity in the equation of state pertaining to that observable. Imagine a multidimensional space, each axis of which is assigned to a chosen external parameter. Each point in this space represents an equilibrium state of the substance. The phase transition in terms of a particular observable is represented by a surface in this space. The transition is of first order if the value of the observable is discontinuous at the surface. In general, a transition is of the $(n+1)$th order if a discontinuity first occurs in an nth derivative of the observable. A region of the space free of transitions of any kind belongs to one phase of the substance. Usually, it is convenient to consider phases with respect to individual observables; the phases and transitions of different observables do not generally coincide.

Usually, attention is focused on two external parameters, with the other relevant parameters considered to be known and kept constant. Then the space discussed reduces to a plane, and a transition is represented by a line in the plane. A plot of this kind is often referred to as a phase diagram of the substance. A transition line may terminate at a point called the critical point. Phenomena occurring near a critical point are referred to as critical phenomena. Often the observable defining the transition has a temperature variation expressible in terms of $(T - T_c)^\beta$, where T_c is the critical temperature and β is called the critical-point exponent. Such a temperature variation is significant in the consideration of phase transitions. A transition line may change its character, the order of transition, at a point; such a point is sometimes called a tricritical point.

Consider, for example, ferromagnetism, which has been treated in Section VI.D.5. The temperature is a parameter. The treatment applies to cases where a small variation in the lattice dimension has no effect of significance. Under such conditions, the phase diagram has an axis for temperature T and an axis for the applied magnetic field H. Of interest is the magnetization, which is a vector. At low temperatures, the magnitude of magnetization for $H = 0$ is that of spontaneous magnetization, but the magnetization does not have a definite direction. For $H \neq 0$, the magnetization has the direction of

H. At $H = 0$, the direction of the finite magnetization changes discontinuously. The temperature axis (T, $H = 0$) is a first-order transition line for the magnetization. The line terminates at T_c, beyond which there is no spontaneous magnetization. Since the magnitude of magnetization is zero at $H = 0$, the magnetization vector varies smoothly as **H** crosses $H = 0$. The point (T_c, $H = 0$) is a critical point. According to (6.234), the critical-point exponent of magnetic susceptibility is $\beta = -1$.

Now consider antiferromagnetism in the simple case discussed in Section VI.D.5, and take the magnitude of alternating magnetization as the observable to be considered. According to the discussion limited to $H = 0$, the alternating magnetization occurs at low temperatures beginning at T_N. Its value is zero for $T = T_N$. Hence (T_N, $H = 0$) is a point of transition, even though there is no discontinuity in the magnitude of alternating magnetization at T_N. It has been shown that this kind of transition continues along a line in the TH plane and joins a line that continues to the H axis. The latter line is characterized by a discontinuity of the magnitude of alternating magnetization. The point (T_t, H_t) at which the two lines join is a tricritical point.

Investigations of the transitions obviously contribute to understanding of the observable. In the case of certain observables exhibiting transitions, the relationships among conditions of their occurrence serve to disseminate the connections in the physics underlying the observables.

BIBLIOGRAPHY

A. Sommerfeld and H. Bethe, Elektronentheorie der Metalle, in *Handbuch der Physik* XXIV/2, Julius Springer, Berlin, 1933.

N. F. Mott and H. Jones, *The Theory of the Properties of Metals and Alloys*, Clarendon Press, Oxford, 1936.

H. Fröhlich, *Elektronentheorie der Metalle*, Julius Springer, Berlin, 1936.

A. H. Wilson, *The Theory of Metals*, Cambridge University Press, Cambridge, 1st ed., 1936; 2nd ed., 1953.

R. E. Peierls, *Quantum Theory of Solids*, Clarendon Press, Oxford, 1955.

M. Born and K. Huang, *Dynamical Theory of Crystal Lattices*, Clarendon Press, Oxford, 1956.

J. F. Nye, *Physical Properties of Crystals*, Clarendon Press, Oxford, 1960.

J. M. Ziman, *Electrons and Phonons*, Clarendon Press, Oxford, 1962.

C. Kittel, *Quantum Theory of Solids*, Wiley, New York, 1963; second revised printing, 1987.

W. A. Harrison, *Solid State Theory*, McGraw-Hill, New York, 1970.

J. Callaway, *Quantum Theory of the Solid State*, Parts A and B, Academic Press, New York, 1974.

N. W. Ashcroft and N. D. Mermin, *Solid State Physics*, Holt, Rinehart and Winston, New York, 1976.

O. Madelung, *Introduction to Solid State Theory*, Springer Verlag, New York, 1978.

Index